A Community-Based Flood Insurance Option

Committee on Community-Based Flood Insurance Options

Water Science and Technology Board
Division on Earth and Life Studies

Board on Mathematical Sciences and Their Applications
Division on Engineering and Physical Sciences

The National Academies of
SCIENCES · ENGINEERING · MEDICINE

AF902833

THE NATIONAL ACADEMIES PRESS
Washington, DC
www.nap.edu

THE NATIONAL ACADEMIES PRESS 500 Fifth Street, NW Washington, DC 20001

This activity was supported by Contract No. HHSP233201400020B with the Federal Emergency Management Agency (via U.S. Department of Health and Human Services). Any opinions, findings, conclusions, or recommendations expressed in this publication are those of the author(s) and do not necessarily reflect the views of the organizations or agencies that provided support for the project.

International Standard Book Number-13: 978-0-309-37468-2
International Standard Book Number-10: 0-309-37468-5

Cover credit: Cover images courtesy of FEMA and FEMA/Bruce Kinney.

Additional copies of this report are available for sale from the National Academies Press, 500 Fifth Street, NW, Keck 360, Washington, DC 20001; (800) 624-6242 or (202) 334-3313; http://www.nap.edu/.

Printed in the United States of America

Suggested citation: National Academies of Sciences, Engineering, and Medicine. 2015. *A Community-Based Flood Insurance Option*. Washington, DC: The National Academies Press.

The National Academies of
SCIENCES · ENGINEERING · MEDICINE

The **National Academy of Sciences** was established in 1863 by an Act of Congress, signed by President Lincoln, as a private, nongovernmental institution to advise the nation on issues related to science and technology. Members are elected by their peers for outstanding contributions to research. Dr. Ralph J. Cicerone is president.

The **National Academy of Engineering** was established in 1964 under the charter of the National Academy of Sciences to bring the practices of engineering to advising the nation. Members are elected by their peers for extraordinary contributions to engineering. Dr. C. D. Mote, Jr., is president.

The **National Academy of Medicine** (formerly the Institute of Medicine) was established in 1970 under the charter of the National Academy of Sciences to advise the nation on medical and health issues. Members are elected by their peers for distinguished contributions to medicine and health. Dr. Victor J. Dzau is president.

The three Academies work together as the **National Academies of Sciences, Engineering, and Medicine** to provide independent, objective analysis and advice to the nation and conduct other activities to solve complex problems and inform public policy decisions. The Academies also encourage education and research, recognize outstanding contributions to knowledge, and increase public understanding in matters of science, engineering, and medicine.

Learn more about the National Academies of Sciences, Engineering, and Medicine at **www.national-academies.org**.

COMMITTEE ON COMMUNITY-BASED FLOOD INSURANCE OPTIONS

HENRY J. VAUX, JR., *Chair*, University of California, Berkeley and Riverside
PATRICIA BORN, Florida State University, Tallahassee
JEFFREY CZAJKOWSKI, University of Pennsylvania, Philadelphia
LLOYD DIXON, RAND Corporation, Santa Monica, California
ROBERT HIRSCH, U.S. Geological Survey, Reston, Virginia
ROGER KASPERSON, Clark University, Worcester, Massachusetts
ROBERT KLEIN, Georgia State University, Atlanta
SANDRA KNIGHT, University of Maryland, College Park
DAVID I. MAURSTAD, OST, Inc., Washington, DC
SALLY McCONKEY, Illinois State Water Survey, Champaign, Illinois
TOMMY WRIGHT, U.S. Census Bureau, Washington, DC
RICHARD ZECKHAUSER, Harvard University, Cambridge, Massachusetts

Staff

ED J. DUNNE, Study Director, Water Science and Technology Board
SCOTT T. WEIDMAN, Director, Board on Mathematical Sciences and Their Applications
BRENDAN McGOVERN, Senior Program Assistant, Water Science and Technology Board

Acknowledgments

The committee appreciates the assistance of the many individuals who provided presentations and additional information to the committee. The committee would like to particularly thank Leonard A. Shabman and Carolyn Kousky, of Resources for the Future, who provided extensive input to the committee.

This report was reviewed in draft form by individuals chosen for their breadth of perspectives and technical expertise in accordance with the procedures approved by the National Academies' Report Review Committee. The purpose of this independent review was to provide candid and critical comments to assist the institution in ensuring that its published report is scientifically credible and that it meets institutional standards for objectivity, evidence, and responsiveness to the study charge. The reviewer comments and draft manuscript remain confidential to protect the deliberative process. We thank the following reviewers for their helpful suggestions, all of which were considered and many of which were wholly or partly incorporated in the final report: Chad Berginnis, The Association of State Floodplain Managers, Madison, WI; Philip Berke, Texas A&M University, College Station; Gerald Galloway (NAE), University of Maryland, College Park; Ed Pasterick, FEMA, retired, Bowie, MD; Bruce Patton, California Earthquake Authority, Sacramento, CA; Mark V. Pauly (IOM), University of Pennsylvania, Philadelphia; and Shannon S. Van Zandt, Texas A&M University, College Station.

Although these reviewers provided many constructive comments and suggestions, they were not asked to endorse the conclusions and recommendations nor did they see the final draft of the report before its release.

The review of this report was overseen by Lawrence D. Brown, University of Pennsylvania, Philadelphia and John Boland, Johns Hopkins University, Baltimore, MD. Appointed by the National Research Council (NRC), they were responsible for making certain that an independent examination of this report was carried out in accordance with institutional procedures and that all review comments received full consideration. Responsibility for the final content of this report rests entirely with the authoring committee and the NRC.

Contents

Summary

River and coastal floods are among the nation's costliest natural disasters. One component of the nation's approach managing flood risk is the availability of flood insurance policies, which are offered on an individual basis primarily through the National Flood Insurance Program (NFIP). Established in 1968, the NFIP is administered by the Federal Emergency Management Agency (FEMA). About 5.4 million individual policies are in the NFIP. Throughout the years, the program has experienced a mixture of successes and persistent challenges. Successes include a large number of policyholders in many flood-prone areas across the nation, the insurance of approximately $1.3 trillion in property, and the majority of policyholders (80 percent) paying NFIP risk-based rates. NFIP challenges include large program debt (roughly $23 billion), relatively low rates of purchase (takeup) in many flood-prone areas, a host of issues regarding affordability of premiums, and a large number of properties that experience severe repetitive flood losses.

To help address some of these challenges within the NFIP, the U.S. Congress passed the Biggert-Waters Flood Insurance Reform Act of 2012 (BW 2012) and the Homeowner Flood Insurance Affordability Act of 2014 (HFIAA 2014). Part of the conversation regarding NFIP reform is the prospect of a community-based flood insurance (CBFI) option—that is a single insurance policy for an entire community. Section 23 of the HFIAA 2014 legislation asked FEMA to evaluate the prospects for CBFI. To satisfy this congressional request, FEMA asked the National Research Council (NRC) to convene an expert committee on a CBFI option (see Statement of Task in Box 1-1, Chapter 1).

This report identifies a range of key issues and questions about a CBFI option that merit further analysis and consideration. Consistent with its statement of task, the committee provides "food for thought"—but does not offer any recommendations—to FEMA and the U.S. Congress as they weigh the possible benefits and challenges of a CBFI option. To have an impact, a CBFI—either as a stand-alone policy option or as part of a suite of policies—will need to address specific challenges.

This report was written on a relatively short timeline and was based in part on presentations and discussions at the two meetings of the National Research Council (NRC) *Committee on Community-based Flood Insurance Options*. The committee's first meeting was held in January 2015 and featured guest presentations; a second meeting was in March 2015 and was convened in a workshop-type format with multiple panels and numerous discussants. (Appendix A includes a list of guest speakers.) Short summaries and key points of these discussions and presentations are included in the body of this report.

To summarize the broad spirit of this report, CBFI may create new opportunities to reduce flood losses, and may enhance the likelihood of communities paying more attention to flood risk mitigation. At the same time, although CBFI may provide a solution in certain circumstances, it is unlikely that it will provide the sole solution for the nation's pressing flood insurance challenges.

WHAT IS COMMUNITY-BASED FLOOD INSURANCE

Community-based flood insurance has been discussed as a possible policy option. A CBFI option could be based on aggregating the dollar sum of flood loss risks for community structures. Costs could be distributed in a variety of ways. For example, one method could be to distribute premium cost according to each individual's assessed flood risk. The community could purchase the insurance, and premiums could be collected using mechanisms such as property taxes or utility charges.

FEMA defines a community as a "political entity that has the authority to adopt and enforce floodplain ordinances for the area under its jurisdiction." In addition, it has the authority to enact and enforce development regulations. This present report does not define a community or CBFI; rather, it cites pertinent past work and examines the future prospects for CBFI—whatever form it may take—and discusses topic areas that will require further evaluation by FEMA and others. Clear definitions will be important if CBFI is to be implemented, and those definitions will be made based on input from elected officials, FEMA, and citizens. It may be helpful to broaden the current FEMA definition of community; a town or city would clearly qualify. It is unclear whether a geographic area in a city, a gated community, or a business district would qualify; however, these entities generally lack authority to enact land use controls to impact floodplain management, building code enforcement, and land use decisions at the local level.

FLOOD RISK MANAGEMENT AND A PRECEDENT FOR A COMMUNITY-BASED FLOOD INSURANCE

Within the NFIP, flood insurance is primarily offered to individuals. Although there is no present CBFI option, communities can participate in the NFIP in several ways. Communities play a substantial role in flood risk management and they often qualify for or receive federal resources to undertake mitigation. Chapter 3 highlights some of these examples to illustrate the precedent for community-based efforts to do mitigation and reduce flood risk.

An excellent example of community-level participation in the NFIP is the Community Rating System (CRS). This is a voluntary, incentive-based program for communities already in the NFIP that undertake various floodplain management measures to prepare for—beyond minimum NFIP requirements—and mitigate against flood losses. The CRS is a class rating system that recognizes measures to help reduce exposure to floods. Taken at the community level, these measures can result in flood insurance premium discounts for policyholders. As of March 2014, about 1,300 communities were actively participating in the CRS. Although this is a small percentage of the 22,000 NFIP participating communities, greater than 65 percent of NFIP policies are written in CRS communities. Within the NFIP, relevant community-based experiences already exist and merit assessment if a CBFI option is pursued.

KEY FLOOD INSURANCE TOPICS

A wide range of topics are pertinent to the NFIP's current challenges. Chapter 3 focuses on key flood insurance topics, which include the following:

- **Participating Communities.** Many different types of communities participate in the NFIP. Some communities have thousands of policies-in-force, while others have only a few, which suggest that one size does not fit all and that CBFI should allow for various approaches.
- **Takeup Rates.** Low takeup rates for NFIP insurance policies, and low rates of policy retention.
- **Community Involvement in Flood Insurance.** Municipal governments do not write NFIP policies. They are, however, involved in floodplain management and mitigation activities that may impact flood insurance premiums.

- **Communication of Flood Risk.** Informed property owners are often well positioned to manage flood risk, which includes the flood insurance purchase decision. FEMA has spent considerable effort, with measurable success, in communicating flood risk at the individual and community levels.
- **Flood Insurance Pricing.** Pricing in the NFIP differs from insurance pricing in the private sector. If discounted or subsidized policyholders had to pay risk-based premiums, then policyholders may be more motivated to engage community officials to undertake measures to reduce flood exposure, thereby reducing flood insurance premiums.
- **Socioeconomic Factors.** Because increasing takeup rates is a goal of the NFIP, it would be helpful if nonfinancial and intuitive thinking be considered. Movement toward full-risk premiums in a community could significantly affect real estate and local tax base. Socioeconomic factors are not limited to private property within a community; other important considerations include public infrastructure and the environment.

RATIONALE AND JUSTIFICATION FOR COMMUNITY-BASED FLOOD INSURANCE

The committee developed a conceptual rationale for CBFI (Chapter 4) that helps to identify where CBFI might be desirable, and where it may not. The Coase Theorem, developed in the economics field, holds that where parties—both individuals and groups—account for all costs and benefits, markets are functional, information is freely and widely available, transactions costs are zero, and economically efficient outcomes are reached irrespective of with whom the property rights are vested. If the collective economic interests of communities and residents fully coincide and are fully accounted for, the outcomes of a flood insurance purchase decision do not rely upon which party—communities or residents—bears responsibility for insurance.

In considering practical applications of the Coase Theorem to flood insurance, there are at least eight reasons (see further details in Chapter 4) to explain why a "Proposition of Responsibility of Insurance is Irrelevant" may fail to hold. Two of these reasons are *free riding*—when some or all residents do not buy insurance because they expect disaster relief to provide adequate post-flood aid—and *externalities*—when self-interested parties fail to account for all of the impacts, when deciding to buy or not buy insurance. These reasons help guide the identification of circumstances when CBFI may be superior or inferior. Choosing the CBFI option requires confidence that insuring at the community level will work better than at the individual level.

There are circumstances where CBFI may provide partial solutions to NFIP challenges. Solutions include reducing administrative and transaction costs, increasing takeup rates, and promoting flood mitigation and floodplain management. Even if CBFI does not effectively address these challenges, it could help in certain areas. For example, moving insurance to the community level would likely enhance attention to risk-reduction activities at that level. Under certain circumstances, however, CBFI may be difficult to implement, for example when a community is not interested or lacks the capability to implement CBFI. Successful implementation would likely require communities to enact some land use restrictions, adopt complementary flood risk management measures and raise additional revenue to pay CBFI premiums.

Central to the concept of insurance is protection from losses incurred from uncertain events such as fire, automobile accidents, and floods. Regarding floods and flood insurance, past patterns of climate and hydrology are limited predictors of future patterns. In addition, changes in land use and population sizes influence flood risk and flood damages in uncertain ways. Scientific evidence shows that flood losses are mostly explained by what is or is not done to the landscape; therefore, efforts to improve landscape management are important. For insurance purposes, uncertainties would need to be evaluated on a case-by-case basis.

DESIGN FEATURES AND CONSIDERATIONS

Chapter 4 of this report has two main sections—a conception of CBFI, and CBFI design considerations. The chapter identifies overarching features and considerations that would require further assessment when planning for and designing CBFI. Each point noted below is elaborated upon in Chapter 4.

- **Risk Bearing and Sharing**

  A CBFI option could conceivably shift risk-bearing to communities, private insurers, or individuals depending on how it is structured. Although as a federal entity, the NFIP may be well positioned to bear the risk, movement to a CBFI option allows for reexamination of how some risk might be transferred to and/or shared by other stakeholders.

- **Responsibilities for Writing Policies and Loss Adjustments**

  Write-your-own (WYO) insurance agents write policies and collect premiums under the NFIP, but CBFI policies would have to be written at the community level. If CBFI were to proceed, FEMA would need to expand a range of administrative duties to process applications from communities.

- **Coverage Limits, Standards, and Compliance**

  Under CBFI, deductible choices would depend on the community's size, the nature of the risk (e.g., type of flooding), existing infrastructure, and other community characteristics. A CBFI option may provide an opportunity to reconsider flood exposure.

- **Underwriting, Pricing, and Allocation of Premium Costs**

  Several complex issues fall under this topic: the extent of actuarial principles to be used in setting premiums (NFIP premiums are legislatively and administratively constrained; see NRC, 2015); the extent to which catastrophic losses would be reflected in premiums for a given community; and the allocation of premium costs among property owners (and renters) in a given community. This third issue could involve deriving some portion of funding from owners of properties that are not in areas subject to flooding.

- **Administrative Capabilities**

  If insurance contracts remain the vehicle for transferring risk, private insurers likely would remain as efficient entities for handling their administration. Communities (probably regardless of definition) would likely not have adequate expertise for undertaking the administration of policies. However, some definitions of community include entities that could effectively and efficiently collect the revenue needed to pay for a community policy through special assessments, property taxes and other sources.

- **Confirming Compliance with Mandatory Purchase Requirements**

  Currently, the mandatory purchase of flood insurance is only for properties that have a federally backed mortgage. A bundled community-based policy that provides a minimum required coverage would need to maintain some aspect of individual coverage insurance and monitoring. This could be administratively burdensome. CBFI could cover all structures to a defined limit. Another alternative would be CBFI that provides a set base coverage amount.

- **Pricing Expertise, Including Valuation of Mitigation Measures**

  If private insurers were to underwrite risks associated with a CBFI policy, then they would want to price polices according to actuarial principles. Private insurers thus would have to possess or acquire the information and expertise to price community-based policies to reflect the risk underwritten and the savings expected from mitigation measures. If the NFIP were to assume the risk of community-based policies, then presumably it would also assume the function of pricing these policies to account for the savings expected from mitigation measures. FEMA has expertise in setting premium costs based on flood risk, and would have to work with communities to communicate individual property coverage costs bundled into a community policy.

1

Introduction

Floods are one of the nation's most destructive and costly natural hazards. Floods are also widespread across the nation; from 2010 to 2015, all 50 U.S. states experienced flooding (FEMA, 2015a). During the 2000s, annual flood damages increased dramatically to nearly $10 billion (ASFPM, 2013). The nation experienced disastrous coastal flooding from hurricanes Katrina in 2005 and Sandy in 2012. The National Flood Insurance Program (NFIP) payouts for the damages caused by the two hurricanes totaled roughly $18 billion and $9 billion in flood losses, respectively (FEMA, 2015b). Since 1978, the NFIP has paid more than $60 billion (in 2012 dollars) in flood losses. The historical means of combatting the adverse impacts of floods involved the construction of public works—dams, levees, flood walls and bypass channels. For many decades, however, the limits of these structural approaches to flood risk management have been recognized. A 1966 report from a special task force on national flood policy, for example, noted that despite the investments of substantial and increasing levels of public funds to construct and operate such projects, flood damages continued to rise (Task Force on Federal Flood Control Policy, 1966).

The Task Force report recommended the use of a broad array of options, including flood insurance, floodplain zoning and floodproofing, to lower the high costs of flood disaster assistance and to attenuate levels of flood risk and exposure. Following publication of the report, Congress established the National Flood Insurance Program (NFIP) in 1968. Originally administered by the U.S. Department of Housing and Urban Development (HUD), the program is now administered by the Federal Emergency Management Agency (FEMA). The NFIP has three main components: mapping, floodplain management, and insurance. Through these components, the program identifies those areas at risk of flooding, aims to reduce flood impacts through mitigation and floodplain management, and offers federal backed flood insurance to protect against flood losses (Hayes and Neal, 2011; GAO, 2013). FEMA's Federal Insurance and Mitigation Administration (FIMA) manages the NFIP.

Since its inception, the NFIP has achieved success and has faced some substantial challenges. Successes include the following:

- Premiums: In fiscal year 2014, the NFIP collected about $3.8 billion in premiums and insured about $1.3 trillion in property (GAO, 2015).
- Flood losses: The NFIP saves the nation an estimated $1.6 billion annually in avoided flood losses (FEMA, 2011a).

Program challenges include the following:

- Financial debt: Through December 2014, FEMA owed the U.S. Treasury $23 billion (GAO, 2015).
- Repetitive loss properties: Between 1978 and 2011, about 166,000 repetitive loss properties had 496,000 claims paid, costing the program fund $12.1 billion (Congressional Research Service, 2013). Furthermore, repetitive loss properties comprised about 1 percent of insured properties, but accounted for 25-30 percent of flood claims. At the time, this cost the NFIP $200 million per year (FEMA, 2009).
- Takeup rates: The Congressional Research Service (2013) cited a news article that suggested that only 15-25 percent of at-risk properties in the Northeast special flood hazard area (SFHA) have flood insurance (Lee, 2012). An earlier, regional estimate in the Midwest suggested a takeup rate of about 20 percent (Galloway, 1995).
- Affordability of premiums: The Biggert-Waters Flood Insurance Reform Act of 2012 (BW 2012), called for movement toward an insurance program with NFIP risk-based premiums. After the Act went into effect, there were many concerns about the financial burden that these increases would place on policyholders (New Orleans Times Picayune, 2013; NRC, 2015). The Homeowner Flood Insurance Affordability Act of 2014 (HFIAA 2014) was enacted to address affordability concerns. HFIAA 2014 limited affordability relief by repealing or altering some BW 2012 requirements (GAO, 2015). For example, it reinstated grandfathering (NRC, 2015) and introduced an annual premium surcharge.
- Subsidized premiums: As of September 30, 2013, the NFIP had more than 1.1 million subsidized flood insurance policies—about 21

percent of all flood insurance policies (GAO, 2014b), which is often cited as being a financial burden on the program.

The NFIP is also challenged by widespread assumption that, because it offers insurance policies and shares some features with private insurance that it operates like private insurance, without the need for federal financial support. Indeed the NFIP has features similar to a private insurance program, but has objectives of ensuring reasonable insurance premiums, having NFIP risk-based premiums, securing high takeup rates, and earning premium and fee income to pay for claims and expenses (NRC, 2015). The NFIP's unique suite of objectives, however, can lead to tradeoffs among its various objectives, which can be difficult to reconcile.

WHAT IS COMMUNITY-BASED FLOOD INSURANCE

Currently, the NFIP offers flood insurance only to individual policyholders. Community-Based Flood Insurance (CBFI) is being investigated as a possible policy option. To some extent, CBFI has already been evaluated as one option among a larger suite of NFIP policy reform options (e.g., FEMA, 2011b; Keybridge Research, 2011). During these earlier evaluations, FEMA (2011b) stated that a community-wide flood insurance premium would be determined by "aggregating the dollar sum of all the individual risk assessments conducted on structures throughout that community." Additionally, it was assumed that premium costs would be distributed according to an individual's assessed flood risk and that the policy would be purchased by the community on behalf of its residents. Flood insurance premiums would be collected through mechanisms such as property taxes or utility-like payments (Congressional Research Service, 2013).

In a more recent study (DePue et al., 2014), CBFI was characterized as the following:

- a governmental or quasi-governmental entity that pays a premium,
- the entity having an opportunity to lower insurance costs through mitigation and other measures,
- insurance costs that may be distributed according to property taxes (for example), and
- the individual cost relative to the total cost of a community-wide premium to be determined by the individual contribution to community risk.

The committee did not offer its own specific definition of CBFI, which is more appropriately in the purview of elected officials, federal and state agency staff, and citizens. The committee viewed CBFI in broad terms to avoid premature narrowing of the discussion. Therefore, this report examines the future prospects for CBFI—whatever form it may take or whatever definitions may be employed.

REPORT FOCUS AND AUDIENCE

This consensus report is the collective effort of a 12-person National Research Council (NRC) committee. This report follows a style that differs slightly from the typical NRC consensus report. This report presents a discussion of theoretical policy options, rather than detailed recommendations for implementing a CBFI option. Much of the report represents the committee's discussion and collective expert judgment. Consistent with its statement of task (Box 1-1), the committee identifies and examines the future prospects of a CBFI option. It neither advocates nor discourages CBFI, nor does it advise on how a CBFI program may be implemented. As

BOX 1-1
Statement of Task and Study Goals

An ad hoc committee of the National Research Council (NRC) will issue a consensus report examining future prospects for community-based flood insurance policies for the United States. Given the lack of experience with community-based flood insurance in the U.S., the committee's report will identify and discuss topic areas and questions that it concludes will require further evaluation—and explain why—in order for FEMA and others to better evaluate strengths and weaknesses of community-based flood insurance.

Examples of these topic areas include

- implementation and feasibility challenges of community-based flood insurance;
- possible terms of community-based flood insurance policies (e.g., options for portions of communities to be covered; renters vs. owners insurance; limits and deductible policies);
- pricing considerations, including possible catastrophic flood losses; and
- potential roles for the private sector.

The committee's report and discussions will consider analogues and lessons from past experiences in the National Flood Insurance Program (NFIP) with pros and cons of individual homeowner policies; relevant information and experience from private sector insurance firms that provide protection against losses in non-flood sectors (e.g., earthquake and fire), and insurance to municipalities; and other information as the committee sees fit.

described herein, CBFI would likely entail both strengths and weaknesses. For example, such an approach has the potential to reduce flood exposure, but at the same time poses some unique implementation and administrative challenges (FEMA, 2011b; Keybridge Research, 2011).

FEMA is the primary audience for this report. Other entities with interest in the topic are the the U.S. Department of Housing and Urban Development and other federal agencies; Congress and congressional staff; local and state officials; flood-prone communities; and private-sector flood insurance companies.

ORIGINS OF THE STUDY

To address many of the NFIP's challenges, the U.S. Congress passed the BW 2012, key provisions of which required raising flood insurance premiums to reflect flood risks, thereby making the program more financially stable. HFIAA 2014 was later signed into law and called for modifications to the BW 2012 Act. For example, HFIAA 2014 restored grandfathered rates and limited yearly increases of flood insurance premiums (GAO,

BOX 1-2
Pertinent Section in the Homeowner Flood Insurance Affordability Act 2014

SEC. 23. STUDY OF VOLUNTARY COMMUNITY-BASED FLOOD INSURANCE

(a) Study—
(1) Study required.—The Administrator shall conduct a study to assess options, methods, and strategies for making available voluntary community-based flood insurance policies through the National Flood Insurance Program.
(2) Considerations.—The study conducted under paragraph (1) shall
(A) take into consideration and analyze how voluntary community-based flood insurance policies
(i) would affect communities having varying economic bases, geographic locations, flood hazard characteristics or classifications, and flood management approaches; and
(ii) could satisfy the applicable requirements under section 102 of the Flood Disaster Protection Act of 1973 (42 U.S.C. 4012a); and
(B) evaluate the advisability of making available voluntary community-based flood insurance policies to communities, subdivisions of communities, and areas of residual risk.
(3) Consultation.—In conducting the study required under paragraph (1), the Administrator may consult with the Comptroller General of the United States, as the Administrator determines is appropriate.

2014a). Other changes required a new surcharge mandated on all policies, and a new requirement to designate a new flood insurance advocate. Section 23 of HFIAA 2014 (Box 1-2) required FEMA to assess options for making voluntary community-based flood insurance available through the NFIP. Shortly after enactment of HFIAA 2014, FEMA asked the NRC to convene an expert committee to prepare a consensus report on the future prospects for a CBFI option.

REPORT ORGANIZATION

Chapter 2 summarizes the sponsor and guest speaker presentations that were made during a public committee meeting on January 7, 2015. Chapter 2 also summarizes presentations from invited speakers—from federal, state, regional, private, and nonprofit sectors—during a workshop on March 30, 2015. Two panel discussions occurred during the workshop: (1) community approaches to flood insurance and (2) flood insurance, risk, and management from the community perspective. These information gathering sessions provided useful input to the committee's deliberations and are reflected in the report's contents. Chapter 3 describes flood risk management strategies, existing community-based options for managing flood risk, and key flood insurance topics. Finally, Chapter 4 has two main sections—(1) a conception of CBFI, and (2) design considerations for CBFI. Chapter 4 discusses past definitions of a community and CBFI, the solutions that CBFI may provide and the challenges it may face, and considerations for the design of a future policy option.

2

Workshop Topics and Presentations

The NRC Committee on Community-based Flood Insurance Options met on January 7-8 and March 30-31, 2015, with both meetings convened in Washington, D.C. During an open session on January 7, the committee heard from the study sponsor and guest speakers. The open workshop on March 30 had two panel discussions: (1) community approaches to flood insurance and (2) flood insurance, risk, and management from the community perspective. The January and March 2015 meetings were convened to examine future prospects for community-based flood insurance (CBFI) and, more specifically, to identify and discuss the benefits and challenges of such an option.

This chapter describes both meetings and considers much of what was presented as analogues and lessons learned from past experiences. These past experiences were shared by representatives from federal, state, local and private entities involved in flood insurance and the broader insurance sector.

PRESENTATIONS AT JANUARY 2015 COMMITTEE MEETING

During the January 2015 meeting, Brian Willsey and Andy Neal from FEMA, Washington, DC presented background information on the National Flood Insurance Program, and FEMA's interest in CBFI. Leonard Shabman and Carolyn Kousky from Resources for the Future (RFF) in Washington, DC discussed CBFI, as well as another NRC study that was at that time under way to assess issues of NFIP premium affordability.[1] David Maurstad from OST Inc. in Washington, DC presented material on defining community-based flood insurance and key topics that merit consideration.

Brian Willsey from FEMA provided an overview of how communities presently participate in the NFIP. To describe the Community Rating System (CRS), he used a case study from Prince George's County, Maryland. Communities that participate in the CRS and are within the special flood hazard area (SFHA), the individual property owners are eligible for a reduced flood insurance premium of up to 45 percent.

Andy Neal from FEMA discussed NFIP reform efforts. He noted that during 2009, FEMA conducted a series of listening sessions with stakeholders, which were Phase I of a three-phase approach to reform efforts. Phase II established an analysis framework, and Phase III evaluated NFIP policy alternatives. One of these alternatives/options was CBFI, which was covered in some detail in the Phase III report. He outlined that detail in a presentation prepared by Keybridge Research for the NFIP reform working group (Keybridge Research, 2011). The Keybridge report cited pros and cons for CBFI that relate to flood exposure, costs borne by individuals, political acceptability, and administrative feasibility.

Carolyn Kousky and Leonard Shabman from RFF discussed their views on CBFI and some ongoing relevant research that they are undertaking. They discussed some pertinent background, the potential for community insurance and some policy design options that they had been studying. Leonard Shabman explained that the NFIP was not designed to be an insurance program only, but rather an assistance program with insurance components for assisting with post-flood recovery, encouraging community floodplain management, and increasing awareness of flood risk. Community roles within the NFIP include adopting flood risk prevention and reduction efforts, cooperating in flood risk mapping and participating in the CRS. Kousky noted that community insurance has been discussed at length, but has not been subjected to a detailed evaluation. Kousky and Shabman defined a community insurance policy as "a flood insurance policy purchased by a community on behalf of its citizens that provides coverage for a specified group of structures." The potential benefits of such an approach to CBFI include a financial incentive, increased resiliency, potential to increase revenues, and lower premiums for individual properties. At the time of the workshop, RFF was undertaking a study funded by the New York Community Trust to evaluate community insurance and produce a report with one or more design questions. Some broad

[1] The NRC Committee on Affordability of National Flood Insurance Program Premiums issued its first report in March 2015 and the committee's second report will be issued in late 2015.

issues being considered included community implementation and feasibility, terms of implementation, and the pricing of a community policy.

David Maurstad from OST Inc. (and a member of the NRC committee), coauthored a presentation entitled "Community-based Flood Insurance: Impacts on the Flood Hazard Management Cycle."[2] He defined CBFI as "a governmental or quasi-governmental entity that pays a premium to insure." Examples include individual homes and community infrastructure. The purchasing entity has the ability to lower insurance through active mitigation and higher deductibles. Maurstad elaborated on how a community-wide insurance might be funded—citing property taxes and utility fees not unlike a stormwater fee as examples. He also identified key topics that affect the implementation of CBFI, which include hazard identification, mitigation, management, insurance, and disaster recovery.

At the March 2015 meeting, the committee held two panel discussions of seven invited experts (see Appendix A) representing federal, state, local and private entities that work with flood insurance in a variety of capacities. The first of these workshops focused on community approaches to flood insurance, while the second focused on flood insurance, risk, and management from the community perspective. Each of the seven panelists made presentations in each of the workshops, during which they shared information or perspectives on flood insurance and addressed one or more of the following questions:

- What is community-based flood insurance (CBFI)?
- What features might a CBFI option have?
- What effect might CBFI have on flood risk management?
- How do communities presently participate in the NFIP?
- How would CBFI be integrated into the CRS?
- What is a community?
- How might a CBFI be priced?
- Does CBFI have the potential to reduce flood exposure?
- Is there a market for CBFI?

The remainder of this chapter summarizes the presentations and pertinent comments from each of the panel discussions.

PANEL DISCUSSION I: COMMUNITY APPROACHES TO FLOOD INSURANCE

André McDonald of the Fort Bend Flood Management Association from Texas began with a brief historical overview of flooding and efforts to

[2] The presentation was initially presented by Michael DePue of Atkins at an Association of State Floodplain Managers conference on June 1-6, 2014. See DePue et al., 2014.

provide flood insurance in the United States. He identified five issues for the future, which should be addressed when devising effective CBFI: (1) What fundamental scientific knowledge is available to guide the development of such a program; (2) Mounting and supporting a program to provide comprehensive floodplain mapping will be crucial; (3) How much flood protection is really needed; (4) How much flood protection can we afford; and (5) Who is best equipped to provide flood insurance coverage.

Bill Lesser from FEMA in Washington, DC, noted that many communities have participated proactively in the existing CRS program to accomplish more than the minimum required floodplain management objectives. He mentioned that moving to a state-based or community-based program may require the involvement of state insurance commissioners who could play an important role(s). He discussed, at some length, the challenges associated with setting premiums for structures built before Flood Insurance Rate Maps (pre-FIRMs) were issued. Mr. Lesser suggested that new policy options may help address a range of problems that arise from changes in flood frequency and magnitude of floods. He emphasized that uncertainty should be accounted for in developing a community-based option and that this would require some emphasis on program flexibility. He drew attention to flood insurance pricing and indicated that the ways in which communities chose to assess fees would crucially impact CBFI success or failure. Stakeholder involvement in the design, implementation, and operation of CBFI will be important. Also important will be the identification of appropriate levels of flood loss reductions.

Bill Nechamen from New York State Department of Environmental Conservation, Albany, focused on the importance of accurate flood mapping (Box 2-1). He noted examples in which people tried to influence the drawing of flood maps to minimize their flood insurance premium costs. This practice tends to adversely impact flood insurance rate setting, because flood insurance premiums require accurate data. In addition, he suggested that FEMA be more flexible in designing and enforcing incentives and regulations. He emphasized the importance of flood mitigation, which must be tied to the CBFI option. Finally, he pointed to the adverse effect of "free-riders"—people who fail to purchase flood insurance but are compensated for flood damages in large storms.

Bob Sokolove from Bank of America in Charlotte, NC focused on the potential for large-scale defaults, in which people simply abandoned and/or forfeit their homes and business because the direct and indirect costs of actual and potential flooding become too high. He noted that if two large storms—with magnitudes similar to hurricanes Sandy and Katrina—struck the United States in quick succession, then there would likely be widespread default in ownership of homes and other buildings (Box 2-1). Similarly, sharp rises in flood insurance premiums, if they result in sharp declines in

BOX 2-1
Major Topics Discussed during Panel Discussion I: Community Approaches to Flood Insurance

- Balancing insurance measures and mitigation measures are important.
- Cross-subsidizing and other forms of subsidizing distort insurance and convey false information to those at risk.
- People focus on ways to avoid or escape flood insurance costs, rather than on the potential benefits of flood insurance.
- Accurate flood mapping and data are important to establishing risk-based flood insurance.
- The pricing of flood insurance is extremely important.
- The capacity of communities to potentially mount and operate CBFI is highly variable.
- If insurance becomes unaffordable and/or catastrophic flood events occur close in time, then widespread defaults on loans and mortgages may ensue.

the value of dwellings or businesses, could lead to widespread defaults. He noted that "green infrastructure" approaches—such as wetland mitigation banking—are examples of novel ways to provide, and perhaps finance, flood protection. He emphasized the importance for all to understand that communities interested in the flood control business should be prepared to do far more than the minimum required.

John Hair from the National Association of Mutual Insurance Companies in Washington, DC, stated that the private insurance industry would likely balk at CBFI for communities that face high risk. He cautioned about practices that entail cross-subsidizing, noting that they may be perceived as unfair and in violation of actuarial principles. The private insurance sector cannot compete with public insurance programs that entail subsidies. He pointed out that some attention should be given to the role of supplementary coverage—privately provided insurance that supplements government-based insurance. He echoed the words of other presenters in noting the importance of integrating insurance with mitigation (Box 2-1).

There is much to learn about CBFI prospects through additional input from the wider private insurance sector. For example, in California several insurers used a commercial loss model that rates fire risk by community, rather than individual structure. Although these types of area maps have been used for years for earthquake risk, their inclusion and increasing sophistication in fire risk is a more recent development. Many of the mitigation issues from a community perspective would be similar for fire and flood (although not earthquake). However, even for earthquake insurance, the issue of overall community resilience is attracting a lot of attention at the state level, particularly in California. In dealing with communities, insurance organizations use a rating system based on the communities' building code enforcement. This practice could provide an important learning analogy for CBFI pricing.

An example from the broader flood insurance sector includes large commercial/industrial flood insurance markets. There is a large commercial/industrial flood insurance market where flood insurance coverage is provided by the private insurance sector. The approach to evaluating and rating an industrial complex, as well as how the private insurance sector works with the client on mitigation and other risk reduction has relevance to a CBFI option.

Katherine Greig from the New York City Mayor's Office stated that effective and comprehensive risk communication is essential to the overall approach to managing flood risk. She noted that the CRS is difficult to apply for in very large and varied communities; therefore, community scale and variety may present several challenges to CBFI. She expressed the view that the issue of low takeup rates probably could only be solved with mandatory flood insurance requirements, which would be politically difficult. She also expressed concern about the possibilities for widespread default.

Vincent Brown from FEMA in Washington, DC, reemphasized the importance of balancing insurance with mitigation. He noted that floods are only one of a number of other natural disasters, such as windstorms and earthquakes. He emphasized the need not only for communicating risk, but also for educating future generations about the nature of natural disasters and for preparing the next generation of flood risk experts and floodplain managers.

PANEL DISCUSSION II: FLOOD INSURANCE, RISK, AND MANAGEMENT FROM THE COMMUNITY PERSPECTIVE

Greig presented the 2007 and 2013 flood insurance rate maps (FIRMs) for New York City to illustrate that floods often go beyond boundaries of the 100-year floodplain. She emphasized the need for effective and ongoing communication about flood risk, noting that preliminary FIRMs are often the best available data. She identified four desirable objectives going forward: (1) reduce risk, (2) improve risk-based pricing, (3) initiate affordability studies, and (4) inform the public.

Brown acknowledged that getting good flood risk information is often difficult to acquire and that it may take several years to update a FIRM.

He emphasized the need to communicate flood risk information early and often, as well as the importance of both insurance and mitigation.

Hair indicated that NAMIC supports the principle of risk-based pricing. He suggested that, to the extent that affordability is an important consideration, the following actions merit further study: performing means testing; augmenting mitigation grants; employing larger deductibles; phasing in rate increases over long periods such as a decade or two, and escrowing insurance payments with mortgage payments.

Sokolove asserted that communities are often reluctant to purchase flood insurance, which impose on them the responsibility for preventing and managing floods. He suggested that flood insurance may fail to achieve the objective of reducing or eliminating disaster assistance because of problems with pricing. Furthermore, many communities lack the capability to mount and execute CBFI because communities are the least capable entities for managing flood insurance. Instead he argued that it would be far better to reform the NFIP, rather than focus on a CBFI option. He emphasized that once a community opted to implement CBFI, it would be very difficult to reverse that decision.

Nechamen also questioned the feasibility of a CBFI option because of a lack of resources at the local level. He recommended that mandatory flood insurance be established for some areas. He stated that there is great uncertainty about what Congress will do when the NFIP expires in 2017. McDonald added that CBFI is just one more option: it will not be applicable in every community, but it should be an option.

The discussion that followed the presentations identified increased resiliency at the community level as an important objective for CBFI. Neverthe-

BOX 2-2
Major Topics Discussed during Panel Discussion II: Flood Insurance, Risk, and Management from a Community Perspective

- There is a fundamental conflict between making flood insurance risk-based, and lowering flood insurance premiums. Flood insurance that accurately reflects risk will usually entail higher premiums.
- The ability and capacity of communities to effectively implement and operate CBFI is highly variable. This means that CBFI may not have universal appeal.
- The importance of effective risk communication to encourage flood insurance purchase—whether individual or community-based—cannot be overstated.

less, the primary motivator of interest in CBFI appears to be the possibility of lower flood insurance premium rates. Some communities are interested in CBFI to increase resiliency through mitigation and to have more widespread insurance coverage.

Box 2-2 highlights the main topics that emerged from this second panel discussion. The discussion also revealed concern about (1) flood damages, and the costs to mitigate and insure against them, are certain to grow; (2) the ability of the federal government to defray these increased costs is in question; and (3) other means to raise the needed revenue are not immediately evident.

3

Flood Risk Management and a Precedent for a Community-Based Flood Insurance Option

Today in the United States, flood insurance is offered through the National Flood Insurance Program (NFIP) via individual policies. Community-based flood insurance (CBFI) is a theoretical concept that has been discussed within the flood risk management community, but has not been implemented as a policy option in practice. There are, however, several programs and opportunities for community-level[1] participation in broader flood risk management, including those through the NFIP.

This chapter describes risk management and how it specifically applies to floods; already existing opportunities for community-level participation as the precedent for community-based options; and key flood insurance topics that include participation and flood insurance takeup rates, community involvement, communication of flood risk, flood insurance premium pricing, and socioeconomic factors.

The basic methods for risk management are applicable to managing flood risk and can be grouped into two categories: (1) risk control and (2) risk financing (Rejda and McNamara, 2014).

Risk control encompasses the methods of avoidance, loss prevention, and loss reduction. Avoidance means that a particular loss exposure is never acquired or undertaken, or an existing loss exposure is abandoned. For example, some flood losses might be avoided by ensuring that no new homes are built in locations where flooding might occur. Moving or abandoning homes in high-risk areas would be another approach to avoiding

[1] FEMA currently defines a community as a political entity that has the authority to adopt and enforce floodplain ordinances for the area under its jurisdiction (FEMA, 2015e).

flood losses. Loss prevention refers to measures that reduce the frequency of losses. In the context of flood risk, loss prevention could include the building and improvement of levees and seawalls and certain other aspects of floodplain management. Loss reduction encompasses measures to reduce the severity of a loss after it occurs. Certain floodproofing methods can be used to reduce the severity of flood losses for a home or building. In some instances, floodproofing measures might prevent a structure from flood damages.

Risk financing includes the techniques of retention, noninsurance transfers, and insurance. It may also include explicit transfers, such as aid. Retention means that the owner of a property (or a community) would retain part or all of the risk associated with flood losses. Noninsurance transfers refer to methods other than insurance by which a risk and its associated financial consequences are transferred to another party. It is unclear how this method could be employed for flood risk in a way that would serve public policy objectives. For example, if a homeowner intends to default on a mortgage because he or she cannot afford or would not plan to repair or rebuild a home damaged by flooding, this could constitute a noninsurance transfer of risk, but this strategy would arguably not be in the public interest. The public interest, in this context, is for housing costs to include the expected loss due to flood risk. Another example that could contribute to a similar result is a lender offering higher mortgage interest rates based on the assumption that a flood will result in loan default. In this situation, the higher mortgage payment is a mechanism to transfer risk. Finally, insurance can be used to transfer flood risk by which a property owner would be partially or fully indemnified for any flood losses by whatever entity or entities that provide the insurance (e.g., the NFIP, private insurance companies). Therefore, when a community considers managing its flood risk, it may consider many of the components described above, with insurance being one mechanism to protect against flood damage.

EXISTING COMMUNITY-BASED OPTIONS FOR MANAGING FLOOD RISK

Communities play a substantial role in flood risk management through a myriad of programs and actions. Community governance structures have unique authorities to administer land use and building codes, levy taxes, collect fees, generate revenue and use other opportunities to reduce or minimize the impacts of flooding. They are often recipients of resources provided by federal programs directed at preparing for, responding to, recovering from, or mitigating flood hazards. Of the programs related to flooding, some are directly linked to the NFIP, while others are tied to disaster and non-disaster programs outside the NFIP.

The following section highlights various ways communities can already participate in the NFIP and serves to illustrate the precedent for a community-based option. These examples include flood risk and floodplain management, the Community Rating System (CRS), the Cooperating Technical Partners Program, and the Flood Mitigation Assistance Program.

Flood Risk and Floodplain Management

Comprehensive flood risk management is a shared responsibility carried out at various levels, from the federal government to the individual. Federal-level activities include building and maintaining structures such as levees, floodwalls and dams; providing real-time flood information and forecasts; communicating flood risk and actions that localities and individuals can undertake to reduce flood risk; preserving or protecting lands that provide for natural storage of floodwaters; and administrating of the NFIP. State-level activities generally include coordinating the NFIP (every state has a designated office for coordinating the NFIP), establishing state building codes and sometimes directly permitting development, coordinating the FEMA hazard mitigation assistance grant programs (every state has a designated State Hazard Mitigation Officer), overseeing dam safety and sometimes levee safety, and undertaking or financing structural measures. At the community level, these same actions are possible, but other options exist through floodzoning and building codes, evacuation plans, and participation in the NFIP, including participation in the CRS.

The NFIP encourages the incorporation of flood hazards into land use decisions. In exchange for the ability of homeowners and businesses to purchase flood insurance from the federal government, communities that join the NFIP agree to adopt and administer the minimum standards, as specified in floodplain federal regulations (44 C.F.R. § 60.3). For example, when a proposal is made to develop within a flood hazard area, federal regulations are intended to minimize risk and flood damages. This means that the community must have the authority to implement and enforce land use and building codes. One NFIP program—the Community Assistance Program - State Supported Services Element (CAP-SSSE)—funds states that provide technical assistance to communities to help them meet their minimum floodplain management ordinances and evaluate their performance.

A 2006 multistate study found that when local governments prepare and implement comprehensive plans for urban development, insured losses from flooding (under the NFIP) are significantly lower than losses sustained by communities that do not adopt such plans (Burby, 2006). This 2006 study recommended several steps that the federal government could take to attract greater attention by local governments to the planning and regulation of urban development and a greater local government role in limiting the adverse financial consequences of hazardous events. Most mitigation policy efforts are isolated and not integrated into the broader activities of local growth management (Lyles et al., 2014). This study demonstrated opportunities for coalition building among stakeholders by seeking ways to produce co-benefits such as habitat protection. Furthermore, the study showed that when mitigation is integrated into comprehensive plans, communities are more likely to adopt and implement regulatory policies aimed at mitigation.

Community Rating System

The Community Rating System is an incentive-based, voluntary program for NFIP communities, with the objective of reducing flood risk.[2] FEMA implemented the CRS in 1990 as an incentive to communities to adopt more rigorous floodplain management strategies and increase flood awareness (FEMA, 2014a). To be eligible, a community must not only participate in the NFIP, but also be in good standing (e.g., no unresolved violations of floodplain management ordinance). The program uses a class rating system from 10 to 1 (1 being best) that offers a 5 percent reduction—and up to a 45 percent reduction—in NFIP flood insurance premiums, with each improvement in class for eligible properties in the Special Flood Hazard Area (SFHA). For eligible properties outside the SFHA, there are only 5 percent and 10 percent discounts. A community accrues credit points to improve its CRS class by engaging in activities such as public information and outreach, mapping and regulations, flood damage reduction, and warning and response. Starting in 2013, the activity of flood insurance promotion was officially added to credit communities that actively encourage residents and businesses to purchase and maintain adequate flood insurance.

Communities can also submit mitigation plans adopted under the Disaster Mitigation Act (DMA) (see Chapter 4 for further discussion on DMA) for CRS credit. A recent study of 71 local mitigation plans in Florida and North Carolina revealed no significant differences between floodplain management elements of hazard mitigation plans that received CRS credit from those that did not receive CRS credit (Berke et al. 2014). Furthermore, among five classes of mitigation policies (land use, structural protection of buildings, flood protection structures, emergency services, and information and awareness), emergency services were given most attention because they were politically the most expedient, did not threaten property values and tax base, and were easiest to administer, while land use policy was given

[2] Multiple CRS resources, such as a coordinator's manual, CRS communities and their classes, number of CRS communities by state, and a national map are available at https://www.fema.gov/national-flood-insurance-program-community-rating-system.

the least attention. The 2014 report concluded that CRS offers promise, but recommended that emergency preparedness and response policies, and investments should not be given credit under the CRS.

Communities expend their own resources to prepare documentation, execute activities, and maintain programs to participate in the CRS; therefore, they have a demonstrated ability to successfully administer a complex flood risk management program. Direct benefits for the community's investment are only offered through premium reductions to the individual policyholders. The premium reductions must be absorbed in the flood insurance policies administered elsewhere. Other benefits to the community include reduced flood risk through mitigation measures such as elevating structures above base flood elevation and retaining floodplain landscapes (Asche, 2014). Because participation in the CRS tends to be in communities that currently have a large number of flood insurance policies, they may also be motivated to participate in a CBFI option.

As of March 2014, about 1,300 communities were participating in the CRS. Although this represents a small percentage of the 22,000 NFIP communities participating in the NFIP, greater than 67 percent of all flood insurance policies were written in CRS participating communities (FEMA, 2014a).

Cooperating Technical Partners (CTP) Program and Risk Map

Another example of how communities participate in the NFIP is the Cooperating Technical Partners (CTP) Program and Risk Map. The CTP Program was developed in 1999 to increase involvement in the NFIP (for example, becoming a more active participant in the flood hazard mapping program) through partnerships with FEMA and regional agencies, state agencies, local entities, tribes, and universities. Main program objectives include maintaining national standards at the local level; provide training and technical assistance; and use data from local sources to help with floodplain management (FEMA, 2014b). There are about 22,000 active NFIP communities, and 240 communities, universities, agencies, and tribal nations have signed agreements with FEMA under the CTP Program (FEMA, 2015c). Under the NFIP flood risk mapping program, there is a NFIP-funded allocation for developing flood insurance rate maps (FIRMs) and a separate, congressionally authorized and appropriated program called the Risk Mapping Assessment and Planning (Risk MAP). Risk Map provides mapping and information to communities to help reduce their flood risk. Under a partnership agreement, roles and responsibilities are established and funding is provided based on eligibilities and agreed-upon activities.

Flood Mitigation Assistance Program

Our final example, which illustrates the precedent of how communities participate in the NFIP is the Flood Mitigation Assistance (FMA) Program. The FMA Program was created in 1994 to reduce the number of insurance claims. It supports mitigation efforts such as floodproofing, as well as rebuilding of property that received significant damage from a severe flood (NRC, 2015). The program provides funds through states, territories, and tribal governments to sub-applicants (communities, tribal agencies, state agencies, tribal governments) who are insured under the NFIP to reduce or eliminate the risk of flood damage. FMA grants can be used to plan projects or to help the grantee manage and administer the program. Funding is provided annually and states can submit projects based on FEMA guidance. The program is a 75 percent federal/25 percent non-federal cost share. In fiscal year 2014, $89 million was available to reduce claims under the NFIP (FEMA, 2014c). In fiscal year 2015, $150 million will be distributed (FEMA, 2015d).

FLOOD INSURANCE—KEY TOPICS

This section discusses several key flood insurance topics within the NFIP: participating communities, policyholder participation and takeup rates, community involvement, communicating flood risk, flood insurance premium pricing, and some socioeconomic factors. Chapter 4 discusses some of these topics in terms of how CBFI may contribute solutions.

Participating Communities

For individual property owners to purchase flood insurance, the community in which the property owner resides must participate in the NFIP (FEMA, 2005a; FEMA, 2005b). The NFIP specifically defines a "community" as a governmental body—including cities, towns, villages, townships, counties, parishes, special districts, states, and Indian nations—with the statutory authority to enact and enforce development regulations (FEMA, 2005a). (See Chapter 4 for further discussion on the definition of community). From a federal perspective, community participation in the NFIP is voluntary. For a community to participate, it must adopt and enforce floodplain management regulations that meet or exceed the minimum requirements set by its state, as well as the NFIP (FEMA, 2005a; FEMA, 2005b). These requirements are meant to ensure that future development in the community will at least meet these minimum requirements and hence better protect the community from flood losses (FEMA, 2005b). In exchange for the implementation and enforcement of these floodplain management regulations, as well as ongoing compliance with the program at large, every property owner in the participating community can purchase flood insurance coverage (FEMA, 2005a). A brief quantitative description of the policies-in-force for NFIP participating communities is presented below.

The following NFIP-participating community data summary is based upon the 2012 NFIP policy database provided to the Wharton Risk Management Center. For the year 2012, within the 50 states, a total of 18,391 NFIP communities were identified from this database encompassing 5.4 million total policies-in-force with $1.25 trillion in insured coverage (building and content) and $3.49 billion of earned premiums.[3]

In 2012, an average NFIP community had 295 policies-in-force with 68 percent of these being single-family residential policies, 22 percent other residential policies (including condominium units), and 5 percent each for multi-family and nonresidential occupancy policies. Across these 295 policies-in-force, an average of $68 million of total exposure ($54.9 million of total building and $13.1 million of total content) was insured in the community. Total earned premiums for this coverage was about $189,600 (see Table 3-1). Thus, for an average policy in an average NFIP community, $231 thousand of total exposure is insured ($186 building and $45 content) with an earned premium of $644 per year.[4]

NFIP communities (Table 3-1) are divided by location in a coastal[5] or non-coastal state. The NFIP has traditionally had higher takeup rates in coastal communities (Dixon et al. 2006; Michel-Kerjan et al., 2012; Kunreuther and Michel-Kerjan, 2013a). Of the total 18,391 NFIP communities, 44% are located in a coastal state, but these communities represent 89 percent of the 5.4 million total policies-in-force. On average, each of the coastal states has 337 active NFIP communities, with 601 policies-in-force on average in each of these coastal communities (67 percent single-family, 5 percent multi-family, 23 percent other residential, and 4 percent nonresidential). Across these 601 policies-in-force, an average of $142.1 million of total exposure ($114.6 million of total building and $27.5 million of total content) was insured in the coastal community. Total earned premiums for this coverage was about $377,200 for the coastal community.

Although non-coastal states have a larger number of NFIP participat-

---

[3] Official statistics provided by the NFIP for 2012 are 5.6 million policies-in-force, $1.29 trillion of coverage, and $3.34 billion of earned premium. See https://www.fema.gov/statistics-calendar-year.

[4] Please note these average values are not split by residential vs. nonresidential types which have different building and content coverage limits—250/100 and 500/500, respectively.

[5] Coastal states include Alabama, Alaska, California, Connecticut, Delaware, Florida, Georgia, Hawaii, Louisiana, Maine, Maryland, Massachusetts, Mississippi, New Hampshire, New Jersey, New York, North Carolina, Oregon, Rhode Island, South Carolina, Texas, Virginia, Washington, and the District of Columbia.

TABLE 3-1 Average Policies-In-Force, Flood Insurance Coverage for Buildings and Contents, and Earned Premiums of NFIP Participating Communities in Both Coastal and Non-Coastal States

	Number of NFIP Communities	Average Policies-in-force	Average Total Building Coverage	Average Total Content Coverage	Average Total Earned Premium
Coastal State	8,076	601	$ 114,557,556	$ 27,531,624	$ 377,209
Non-Coastal State	10,315	55	$ 8,243,244	$ 1,873,052	$ 42,743
Total U.S.	18,391	295	$ 54,928,818	$ 13,140,445	$ 189,616

SOURCE: 2012 NFIP policy database.

ing communities (on average each non-coastal state has 382 active NFIP communities), they support fewer policies-in-force, 55 policies-in-force on average (78 percent single-family, 4 percent multi-family, 7 percent other residential, and 11 percent nonresidential). Across these 55 policies-in-force, an average of $10.1 million of total exposure ($8.2 million of total building and $1.9 million of total content) was insured in the non-coastal community. Total earned premiums for this coverage was about $42,700 for the non-coastal community. Thus, for an average policy in an average non-coastal NFIP community, $183 thousand of total exposure is insured ($149 building and $34 content) with an earned premium of $775 per year.

Figure 3-1 illustrates the complete distribution of the number of policies-in-force in each community split by coastal and non-coastal states. Both community policies-in-force distributions are skewed implying non-symmetry in participating community "size," and thus making it more difficult to determine a typical average community size value as discussed above. In coastal states, 33 percent of communities have 10 or fewer

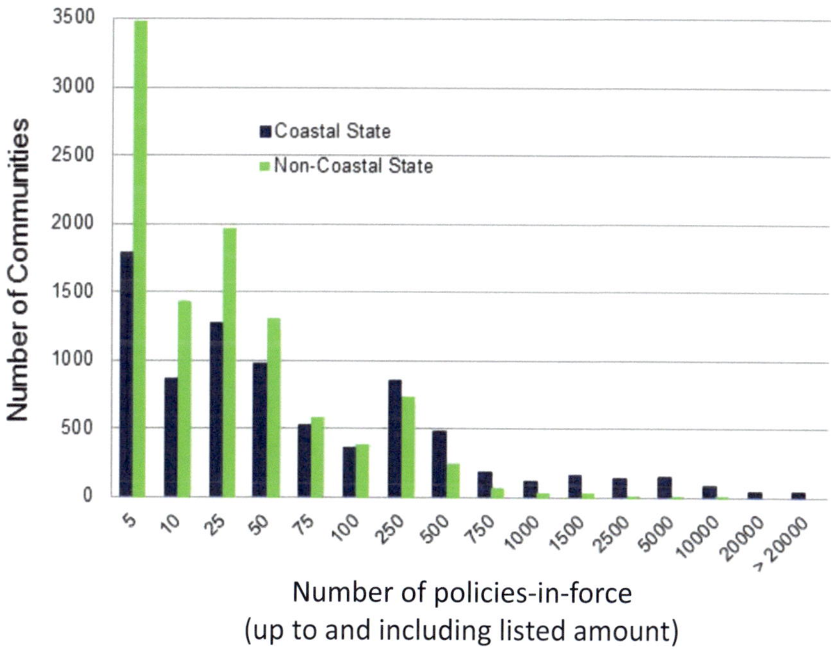

FIGURE 3-1 Number of NFIP-participating communities in coastal and non-coastal states plotted against the number of policies-in-force.
SOURCE: 2012 NFIP policy database provided to the Wharton Risk Management Center.

policies-in-force, with a median value of 27 policies-in-force, and a maximum value of 186,591 policies-in-force for a community in Florida. In non-coastal states, 48 percent of communities have 10 or fewer policies-in-force, with a median value of 12 policies-in-force, and a maximum value of 8,624 policies-in-force for a community in Arizona.

Overall, the data illustrate a flood insurance program that impacts a large number of NFIP-participating communities in different ways. For example, some participating communities have 100,000 or more policies-in-force, whereas a substantial portion (41 percent) of communities in the overall NFIP portfolio has 10 or fewer policies-in-force. Furthermore, this variation in community size and corresponding insurance coverage exists independent of whether a state is coastal or non-coastal. Thus, it is not a simple matter to define an average NFIP community, suggesting it may be complex to develop a one-size-fits-all community-based approach to flood insurance.

Policyholder Participation and Takeup Rates

During the early 1970s, only 95,000 NFIP policies were in place. The NFIP paid only $3 million in claims following Tropical Storm Agnes in 1972, even though total damages were estimated to be $3 billion (Anderson, 1974; FEMA, 2002). In response to this low level of takeup, Congress passed the Flood Disaster Protection Act of 1973, which directed federally regulated lenders to require flood insurance as a condition of granting or continuing a loan on structures in the SFHA (the mandatory flood insurance purchase requirement; see further discussion in Chapter 4). Even though takeup rates increased in general, in areas affected by the Midwest floods in 1993, only about 20 percent of structures were insured (Galloway, 1995). Congress subsequently strengthened the mandatory purchase requirement in the National Flood Insurance Reform Act of 1994. As described below, however, takeup rates still remain low.

About 5.5 million policies were in force as of October 2013. Estimates of takeup rates at the national level are difficult because of a lack of data (NRC, 2015), but estimates based on detailed analysis of individual properties arrive at similar results. Based on a nationwide sample, it was estimated that approximately 50 percent of one-to-four person family homes in the SFHA were insured for flood losses (Dixon et al., 2006). Similar findings were found for one-to-four person homes in New York City on the eve of Hurricane Sandy (Dixon et al., 2013). Of these homes in the SFHA, 55 percent had flood insurance (approximately three-quarters were subject to the mandatory purchase requirement). Thus, even though the proportion of residential structures covered by the program has increased greatly since its early years, roughly half of one-to-four family homes in the SFHA still lack insurance. Takeup rates outside of the SFHA are much lower. Estimates for one-to-four family homes run in the low single digits, with research finding the takeup rate to be less than 1 percent outside the SFHA (Dixon et al., 2006).

Contributing to modest takeup rates is the issue of flood insurance retention. NFIP policies must be renewed annually. A 2012 study examined the number of years NFIP policies remain in place and found that, of the 840,000 new policies issued in 2001, 27 percent had lapsed after 1 year and 51 percent had lapsed after 2 years (Michel-Kerjan et al., 2012). To help address the takeup rate, FEMA expended considerable effort to communicate flood risk through the FloodSmart program, Write-Your-Own (WYO) agents, and community efforts. For example, a specific goal of the FloodSmart program is to increase takeup rates (FEMA, 2015a). The GAO (2010) reported that the NFIP had demonstrated a 24 percent increase in the number of policies between the launch of FloodSmart program in 2004 to the time of this present report. Although promising, NFIP policy takeup rates remain low.

The NFIP has several program objectives, one of which is to encourage the purchase of flood insurance as an alternative to reliance on federal disaster assistance. Hence, for any policy option intended to increase takeup rates, it would be useful to assess how a CBFI option might affect program objectives and whether tradeoffs are involved.

Community Involvement in Flood Insurance

Local governments are typically involved in flood mitigation activities such as designing and building structures for flood damage mitigation; enacting and enforcing floodplain management regulations; establishing and operating flood warning systems and emergency management plans; and regulating land development in upland areas to reduce the extent to which new impervious surfaces on these lands will increase flooding downstream.

When it comes to flood insurance, however, local governments have no formal role, although some of their mitigation activities may enable flood insurance policyholders to obtain reduced premiums through the CRS. As mentioned above, in 2013 the activity of flood insurance promotion was officially added to the CRS to credit communities that actively encourage residents and businesses to purchase and maintain adequate flood insurance. However, it is the policyholders, and not the governmental entity, that reap the direct benefit of a reduced premium from a better CRS score. Of course, the community reaps the indirect benefit of better protection and lower losses from floods. If a CBFI option were available to communities to formally participate in the NFIP, then it would have the direct beneficial effect of making flood hazards and flood hazard reduction important topics of debate in the community. Chapter 4 considers how CBFI might enhance community involvement.

Communication of Flood Risk at the Community Level

It is important for property owners to understand their exposure to flood risk, as well as the consequences of their failure to buy flood insurance, especially if they live in a high-risk area such as the SFHA. Informed property owners are often well positioned to make good choices in managing their exposure to floods. The same can be said for communities that can undertake measures to better manage their floodplains and mitigate flood hazards. The NFIP through its mapping, floodplain management and insurance activities has spent considerable resources to communicate flood risk. For example, the Risk MAP program, which maps flood risk on behalf of the NFIP, conducts outreach to communities in the development and distribution of flood maps, along with activities to engage communities and communicate risks. These actions are intended to support efforts to increase public awareness of flood risk and emphasize the importance of flood mapping data in contributing to appropriate risk management decisions and actions. FEMA and other federal agencies have also implemented a High Water Mark Initiative, wherein communities post signs about their historical flood events and conduct ongoing education to build awareness.

FEMA is engaged in a compreshensive effort to reform the NFIP, which was initiated before the Biggert-Waters Flood Insurance Reform Act of 2012 (BW 2012) was enacted. The first phase involved stakeholder sessions, during which federal, state, local and tribal governments, and other entities shared concerns and recommendations (FEMA, 2009). Similar sessions were held around the country in preparation of the National Mitigation Framework, which focuses on reducing risks, as well as empowering communities to take action and increase community resiliencey (FEMA, 2010a).

Perhaps the most signicant communications effort undertaken by FEMA is the FloodSmart program, mentioned previously. This NFIP-funded advertising and marketing program is directed at encouraging homeowners to buy flood insurance. The program provides information about five main topics: flooding and flood risk, the NFIP, residential flood insurance, commercial flood insurance, and preparation and recovery (NRC, 2013). Through internet sites, television advertising, brochures, and other forms of marketing, the program disseminates information about policies, program changes, and other topics of interest to the consumer (FEMA, 2015a). The importance of communicating risk and the need for insurance suggests that periodic evaluation of communication strategies would be useful.

Flood Insurance Premium Pricing

Many pricing challenges are associated with flood insurance. Subsequently, BW 2012 called for substantial changes to how the NFIP prices policies. More specifically, its provisions required the NFIP to move to risk-based pricing of most properties. However, when affected policyholders voiced their concerns about the implications of BW 2012, the Homeowner Flood Insurance Affordability Act of 2014 (HFIAA 2014) was enacted. HFIAA 2014 limited the premium rate increases that would have occurred for some properties if all of the BW 2012 provisions had taken effect. In particular, HFIAA 2014 delayed, but did not reverse, increases in the rates for pre-FIRM properties, but it reinstated grandfathering (NRC, 2015). The intention was to offset the revenue losses from grandfathered and CRS-discounted premiums by increasing premiums on all policies, but it is uncertain whether this objective has been achieved (NRC, 2015). HFIAA also introduced an annual premium surcharge, which will be deposited into a reserve fund.

Issues that arise from the NFIP's approach to pricing flood insurance include, but are not limited to, financial deficits, distorted incentives with respect to flood risk management, and adverse selection. These pricing topics are discussed below.

Financial Deficits

The NFIP has incurred substantial financial debt in recent years. There are several reasons for this including, but not limited to, NFIP premium pricing practices. Since its inception, the NFIP has incurred catastrophic losses—$1 billion or more—in 7 of its 44 years. Throughout its history, the NFIP has had to borrow money from the Treasury on several occasions when its claims payments and other costs exceeded its revenues and reserves. Until 2005, the program never exceeded its cumulative debt of $1 billion, and was able to repay its loans when its claims payments were relatively low. This changed in 2005 when claim payments resulting from hurricanes Katrina, Rita, and Wilma caused huge increases in its deficits and debt. Since 2005, the NFIP's outstanding debt has risen from $225 million to $23 billion (through December 2014). Although its revenues steadily increased during this period they have fallen far short of total claims payouts. The years 2005 ($17.8 billion) and 2012 ($8.8 billion) were especially costly with respect to claim payouts.[6]

There are differences between how a private insurance company prices its loss exposures and manages its finances, and how the NFIP does so. In the absence of regulatory actions, a private insurer will set prices so that, over the long term, it will generate sufficient revenues to fully cover its costs. In low-loss years it will contribute to its "surplus," from which it can

[6] Data sourced from https://www.fema.gov/loss-dollars-paid-calendar-year [accessed on June 13, 2015].

draw funds in high-loss years. In contrast, the NFIP's pricing structure has not generated sufficient revenues to fully fund its expected or actual losses. NFIP's pricing incorporates other program goals that are at times at odds with the ability to cover payouts for losses, which makes the NFIP fundamentally different from private insurance (Kousky and Shabman, 2014; NRC, 2015). If the NFIP operated like a private insurer, then it would include a loading for a risk-based return on capital in its premiums. The NFIP is not pricing to build up substantial reserves (from an actuarial perspective) to cover payouts in high-loss years and/or purchase reinsurance or catastrophe bonds. BW 2012 does require the NFIP to institute assessments intended to build up loss reserves, but the GAO (2014a) has concluded that these assessments will be insufficient to achieve their intended objective.

A significant contributing factor to the NFIP's debt seems to have been its inability to charge full-risk premiums. For example, the GAO (2014b) estimated the NFIP's net "foregone premiums" during the period 2002-2013 to be between $11 billion and $17 billion. The GAO defined "foregone premiums" as the difference between "subsidized" premiums and "full-cost" premiums (GAO, 2014b).

BW 2012 required the NFIP to begin to build a reserve equal to 1 percent of its total potential loss exposures, but with its current pricing structure, it will not be able to meet this requirement (GAO, 2014a). In 2014, the NFIP added a 5 percent assessment on policies to build this reserve fund. However, it has been estimated that a 25 percent assessment would be needed (GAO, 2014b). The GAO (2014a) concluded that it is not possible for the NFIP to meet its reserve target under the limitations imposed by HFIAA 2014. There also is an issue as to whether the NFIP should be required to use its current rate structure and future revenues to pay off its outstanding debt to the Treasury given the statutory limits on its rate structure. The program was never intended to be actuarially sound; rather, it was assumed that FEMA would have to borrow from the federal Treasury.

Distorted Incentives

The manner in which insurance coverage is priced can have substantial effects on insurance policyholders' incentives to manage their risk. Providing insurance, in and of itself, can lead to moral hazard by diminishing insureds' incentives to prevent or avoid losses. Measures can be taken in the design, underwriting and pricing of insurance policies to mitigate moral hazard. Risk-based pricing is one of these measures. The more that someone pays for insurance that is tied to the risk of loss, the more he or she will be induced to take steps to reduce the risk of loss or cancel the policy.

Individuals and households can control or influence their exposure to flood risk primarily through their decisions about the location of the homes they own (or rent) and measures they can take to make their homes more flood-resistant. In terms of location, individuals and households can choose to build, buy, or rent homes in areas that have a high, moderate, or low propensity for floods. A home and its contents can be made more flood-resistant through measures such as increased elevation, dry floodproofing and wet floodproofing, among others. Similar observations apply to non-residential properties. Communities can also take steps to reduce the flood exposure of properties within their boundaries through floodplain management and hazard mitigation.

If property owners, which have a subsidized flood insurance policy, pay less than what they would pay if there policy was not subsidized, there is less incentive to mitigate against flood loss. As discussed above, there are three primary groups of polices or properties that are currently charged less than what FEMA considers to be full-risk premiums—pre-FIRM subsidized policies, grandfathered policies, and policies that receive CRS discounts. Further, there are 6 categories of subsidized policies provided by NFIP. Some examples of these are pre-FIRM policies, policies on properties newly mapped into the SFHA, and policies within different flood zones.

The subsidies provided to the owners of pre-FIRM subsidized properties likely will not affect decisions on where to build new homes, because they apply to existing properties only. They will, however, affect the decisions of prospective buyers of these properties, because HFIAA 2014 allows new owners of these properties to maintain coverage at the historical rate. Through time, the pre-FIRM subsidized rates will gradually increase, although at a slower pace than what the BW 2012 had mandated.

The premium reductions received by the owners of pre-FIRM subsidized, grandfathered, and discounted CRS properties would also be expected to reduce their incentives to undertake measures to make their properties less vulnerable to flood damage.

If a property owner is paying an artificially low premium, then he or she may perceive that any reduction in their premium to be small relative to the cost of undertaking a mitigation measure such as floodproofing. That said, as the rates for pre-FIRM properties gradually increase, the perceived benefits of floodproofing investments should also increase relative to their costs. The time horizon that a property owner would use in any cost-benefit analysis would be important, because the net present value of the anticipated premium savings over an extended period could be substantial.

Discounts and subsidies received by NFIP policyholders would also be expected to diminish the incentives for communities to manage their floodplains and invest in hazard mitigation. If the owners (or renters) of these properties paid risk-based rates, then they would be more motivated to request officials in their communities to undertake measures to reduce their flood exposure to lower their premiums.

Adverse Selection

If an insurer is unable to charge premiums commensurate with individuals' risk of loss, then it will be exposed to the problem of adverse selection (Box 3-1). Arguably, the NFIP is purposely structured to encourage adverse selection at the high end of the risk spectrum, because one of the NFIP objectives is to motivate owners of high-risk properties to buy flood insurance, even if the premiums paid are less than what it costs to insure. If owners of moderate or low-risk properties perceive (rightly or wrongly) that the premiums they would pay are "too high" relative to their risk of loss, then they will be less inclined to buy flood insurance. This is a problem for two reasons: (1) given that owners of moderate- and low-risk properties still have some flood exposure if they choose not to buy flood insurance, then they risk incurring uninsured flood losses; and (2) if the NFIP pricing structure discourages the owners and renters of moderate- and low-risk properties from buying flood insurance, then the program's ability to achieve a broader pool of risk exposures, which is generally considered to be desirable for other types of insurance, is compromised. Two concepts are pertinent here: (1) increasing the size of an insurance pool can reduce objective based risk, which could be viewed as an economy of scale. However, insurance experts posit that the efficiencies gained through pooling larger numbers of loss exposures does not require that the exposures be of "similar risk" (Harrington and Niehaus, 2004; Rejda and McNamara, 2014); and (2) it would be impractical to establish a pool of insureds with similar

BOX 3-1
Adverse Selection

"Adverse selection is the tendency of persons with a higher-than-average chance of loss to seek insurance at standard (average) rates, which if not controlled by underwriting, results in higher-than-expected loss levels" (Rejda and McNamara, 2014). Adverse selection can occur because of asymmetric information (individuals know more about their risk level than insurers) or government restrictions on insurers' ability to charge risk-based prices, or a combination of both. The extent to which adverse selection becomes a problem can depend upon the severity of the asymmetric information problem, the extent to which people change their insurance purchasing decision as premiums change, and any restrictions on insurers' ability to employ risk-based underwriting and pricing. Also, the degree to which the purchase of insurance is voluntary or compulsory will affect adverse selection. If the purchase of insurance is voluntary, then insurers will be subject to greater adverse selection.

risk levels: insurers generally seek to achieve "a balance" within their pools of insureds so that they are not overly weighted with high-risk insureds.

The extent to which low takeup rate for owners of moderate- and low-risk properties can be attributed to adverse selection is a matter for speculation. In addition, if the owners of some high-risk properties are charged a premium higher than what it costs to insure them, then less-than-satisfactory takeup rates in high-risk areas could result. Given current limitations on the mandatory purchase requirement and the problems that have been encountered in its enforcement, this issue may warrant further investigation. The NFIP's gradual movement to risk-based rates for pre-FIRM structures would be expected to make the program less subject to adverse selection. The assessments and surcharges being imposed on all insureds would be expected to have the opposite effect. These are issues that warrant consideration as further modifications to the NFIP, such as CBFI are explored. However, there are many pricing challenges associated with flood insurance and they are neither created by nor avoided by the use of CBFI.

Socioeconomic Factors

Modest takeup rates for individual property flood insurance exist within NFIP-participating communities in SFHAs. These rates are less than modest in NFIP-participating communities outside of the SFHAs. These modest to less than modest takeup rates can potentially be explained by property owners deliberatively evaluating premiums and deductibles for various flood insurance policies against the probability and severity of flood losses (NRC, 2015), and rationally choosing to purchase little or no flood insurance within this cost/benefit, decision-making context. However, the purchase of flood insurance may not be based solely on rational, financial, cost/benefit considerations, but may be subject to nonfinancial considerations and intuitive (rather than deliberative) thinking (NRC, 2015).

Intuitive thinking is predicated upon emotionally driven responses and mental shortcuts that are often conditioned by personal past experience, social context, and cultural factors (Kunreuther et al., 2014). This type of thinking does not necessarily work well when making decisions around low-probability, high-consequence events such as flooding (Kunreuther et al., 2014), because it is subject to several deliberative biases. For example, homeowners will often purchase flood insurance immediately after a salient flood event they now considered to be most likely (availability bias), but will allow this policy to lapse after a few years if no further flood event occurs (NRC, 2015). Other responses and mental shortcuts include budgeting heuristics, biases in temporal planning (e.g., hyperbolic discounting), learning failures and status quo bias (reluctance to consider new alternatives), risk framing, and social norms and interdependencies (see Kunreuther et al.,

2014, and NRC, 2015 for a more detailed discussion). These nonfinancial and intuitive thinking considerations should be accounted for when designing policy options aimed at increasing flood insurance takeup rates.

To encourage participation in the NFIP, both pre-FIRM subsidized policies and grandfathered policies were instituted to avoid penalizing property owners (and corresponding local communities) who might otherwise see a significant decrease in property values. Furthermore, these policy premiums were neither means tested, nor targeted at lower-income property owners (Kunreuther and Michel-Kerjan, 2013a). It is estimated that 20 percent of the 5.5 million policyholders (34 percent of all policies in the SFHA) pay about 40-45 percent of the true full-risk premium (Hayes and Neal, 2011; GAO, 2013). These values apply to pre-FIRM subsidized policyholders.

Thus, movement toward premiums that reflect full risk for all property owners in a community—ostensibly what would occur under a CBFI option that removes premium subsidies—would ideally account for the premium increases potentially imposed, especially on lower income property owners for whom this would be an financial burden. That is, full-risk premiums for all property owners should address issues of affordability of insurance. A 2015 report provides a detailed example of the magnitude of rate increases for a subset of existing pre-FIRM SFHA policyholders in Charleston, South Carolina moving toward post-FIRM, full-risk premiums based upon the current NFIP rating manual (Zhao et al., 2015).

The broader flood insurance affordability impact will also likely be experienced in the near term because of current issues of modest to low insurance takeup rates and poor enforcement/compliance. These low market penetration issues are further compounded by expectations of disaster relief, whereby property owners in high flood risk areas have not taken steps to reduce their exposure or have adequate insurance in place because they assume they will be compensated if a disaster occurs (Kunreuther and Michel-Kerjan, 2013a). In fact, actual disaster assistance is likely to be much smaller and more uncertain than is commonly perceived. Increasing flood insurance market penetration is unlikely to reduce disaster assistance (Dixon et al., 2006). All told, low takeup rates, poor enforcement/compliance, and disaster relief expectations are believed to have led to continued development in high-risk flood areas without the corresponding purchase of flood insurance.

Equity and economic stability are prominent socioeconomic considerations for a community. Full-risk premiums may impose a significant financial burden on lower-income households, and a number of high-income policyholders have benefitted from NFIP-discounted policies. For example, wealthy property owners with waterfront homes that are not their primary residence have purchased insurance from the NFIP at subsidized rates; however, subsidies are now being phased out on these properties at a rate of

25 percent per year. Although movement toward full-risk premiums would impact these policyholders, the financial burden would not be significant; more importantly, it would add to the community's overall flood insurance portfolio, potentially unequally shared amongst wealthy and non-wealthy property owners through cross-subsidized insurance premiums. As mentioned above, premium discounts were also provided to avoid negative impacts on property values. Thus, movement toward full-risk premiums in a community could have significant real estate and consequently local tax base implications. Decreased property values could impact a community today, as well as into the future, because lower tax revenues would force public spending trade-offs. Again, it is doubtful that these negative impacts would be felt equally across all NFIP participants.

A large body of literature demonstrates that socially vulnerable populations—particularly low-income and minorities—experience greater impacts from storms and flooding (Van Zandt et al, 2012), in part because they are unable or unwilling to adequately insure their property, and in part because they are at greater risk of exposure. Within communities, people and households are not distributed randomly—the risk of exposure is not borne equally. In many communities, low-income households are located in low-lying areas, often in low-quality buildings. These situations result from market forces and intentional development decisions that place lower-income households (including rental housing) in less desirable/more hazardous areas. As a consequence, low-income households may have limited housing choices. Part of the appeal of CBFI is the potential to spread risk throughout the community (Keybridge Research, 2011), which may encourage the protection of vulnerable areas.

Finally, community socioeconomic factors are not limited to impacts on private property owners within a community. Two important considerations in this regard are public infrastructure and the environment. With regard to the former, as part of the 1988 Stafford Act, after the declaration of a Presidential disaster the federal government is authorized to provide at least 75 percent of the funds, that is, disaster assistance/relief, required to restore damage to infrastructure and public buildings (Kunreuther and Michel-Kerjan, 2013b), most of which are uninsured and/or unmitigated toward disaster impacts. FEMA has responsibility for coordinating this disaster relief.

Disaster assistance has not been limited to FEMA allocations. It has been distributed through numerous other federal agencies (Staff, Subcommittee on Economic Development, Public Buildings, and Emergency Management, 2015). For example, $60 billion in total disaster relief has been appropriated due to Hurricane Sandy with the largest allocation ($15.2 billion) going to the Department of Housing and Urban Development Community Development Funds. Other large disaster relief appropriations went to the Department of Transportation Federal Transit Administration Emergency Relief Program and the U.S. Army Corps of Engineers (Staff, Subcommittee on Economic Development, Public Buildings, and Emergency Management, 2015). This $60 billion to provide disaster relief primarily to infrastructure and public buildings dwarfs the $7.8 billion in NFIP claims paid.[7] In light of this, calls have been made for communities in hazard prone areas to pay insurance with full-risk premiums to cover damage to public infrastructure and buildings (Kunreuther and Michel-Kerjan, 2013b), and these issues merit consideration in a CBFI option.

SUMMARY

Within the NFIP, flood insurance is primarily offered to individuals. Although there is no present CBFI option, communities can participate in the NFIP in several ways. Communities play a substantial role in flood risk management, often receiving federal resources to engage in such activities. This chapter highlights some of these programs to illustrate the precedent for community-based approaches. For example, the CRS is an incentive based voluntary program for NFIP communities to reduce flood risk. The CRS uses a class rating system and for each improvement in class, a discount is offered on flood insurance to individual policyholders. Other examples of community participation in the NFIP include flood risk and floodplain management, the Cooperating Technical Partners Program and Risk Map, and the Flood Mitigation Assistance Program. These experiences merit further review if a CBFI option is pursued.

This chapter discussed the following key flood insurance topics:

- **Participating Communities.** The NFIP impacts participating communities in different ways. Some communities have a large number of policies-in-force, while other communities have very few policies-in-force. Therefore, there is little evidence to define an average NFIP community, suggesting that a one-size-fits-all approach to a CBFI option may be difficult to develop.
- **Takeup Rates.** Takeup and retention rates for flood insurance are often low. Mandatory purchase requirements, which might raise both rates, are likely to encounter political resistance.
- **Community Involvement in Flood Insurance.** Municipal governments do not write NFIP policies. They are, however, involved in floodplain management and mitigation activities that may impact flood insurance premiums.

[7] Significant flood events. Available at https://www.fema.gov/significant-flood-events [accessed on July 8, 2015].

- **Communication of Flood Risk.** Informed property owners are often well positioned to manage flood risk, which includes the flood insurance purchase decision. FEMA has spent considerable effort, with measurable success, in communicating flood risk at the individual and community levels.
- **Flood Insurance Pricing.** Pricing within the NFIP differs from pricing in the private sector. If discounted or subsidized policyholders had to pay risk-based rates, then policyholders may be more motivated to engage community officials to undertake measures to reduce flood exposure, thereby reducing flood insurance premiums.
- **Socioeconomic Factors.** Because increasing takeup rates is a goal, nonfinancial and intuitive thinking merit consideration. Movement toward full-risk premiums in a community could significantly affect real estate values and the local tax base. Socioeconomic factors are not limited to private property owners within a community; other important considerations include public infrastructure and the environment.

4

Community-Based Flood Insurance: Issues and Considerations

This report identified and described some of the major problems facing the National Flood Insurance Program (NFIP). These problems include the pricing of flood insurance premiums, takeup rates, efforts to support floodplain management, and socioeconomic factors. In theory at least, a community-based flood insurance (CBFI) option may have the potential to help solve some of these problems. For example, if a community purchased insurance for all dwellings within it, then the takeup problem would be solved.

Concerns exist about flood insurance purchases by individuals. For example, community members, perhaps motivated by altruistic feelings, paternalistic concerns, or by concerns that their lives would be affected by others' uninsured losses, would prefer that their neighbors be insured. The flooding of uninsured properties in the community could have adverse impacts on property values and could generate other adverse physical, social, and economic consequences. Various government entities, most notably the federal government, also secure a positive externality from insurance purchase: covered individuals are less likely to require costly disaster relief.

A major objective of flood insurance is to ensure that both individuals and the community undertake appropriate policies to mitigate the risk of a loss, as well as the size of a loss should a flood occur. In theory, if insurance pricing was full risk and community leaders accounted their constituents' costs (including insurance costs), then this objective would be achieved. Conceivably, if flood insurance premiums were paid at the community level, then leaders would become more strongly oriented toward mitigation efforts on behalf of their constituents. These are all conceptual arguments;

mere identification of their existence does not sufficiently support or oppose significant efforts to promote CBFI. A key requirement would be to evaluate their potential impact should communities purchase flood insurance.

Chapter 4 further discusses some of the flood insurance topics that merit consideration when evaluating the future prospects for CBFI. This chapter has two main sections: (1) a conception of CBFI, and (2) design considerations for CBFI.

A CONCEPTION OF COMMUNITY-BASED FLOOD INSURANCE

A basic premise underlying CBFI is that it may be less expensive and more effective to write single flood insurance policies at the community level than having insurance companies or the NFIP write multiple policies at the individual level. For this to be true, either the community must be more efficient than FEMA in writing policies covering individual entities, or the shift of responsibility from individual owners to the community will foster incentives to protect properties from floods.

The following section discusses the following topics: what is a community; the rationale for CBFI; spillovers across communities; how CBFI might provide solutions; and reasons why CBFI might not be successful.

What Is a Community?

There is no single definition of what constitutes a "community," and many different definitions exist in an array of contexts. For example, definitions of "community" in Webster's Ninth New Collegiate Dictionary include

> "a unified group of individuals"; "people with common interests living in a particular area"; "a group linked by a common policy"; and "a body of persons having a common history or common social, economic, and political interests."

FEMA defines a community as

> "a political entity that has the authority to adopt and enforce floodplain ordinances for the area under its jurisdiction" (FEMA, 2015e).

A recent study (NRC, 2015) modified the FEMA definition and stated

> "A political entity that has the authority to adopt and enforce floodplain ordinances (engage in mitigation) for the area under its jurisdiction as a requirement of the National Flood Insurance Program. In most cases, a community is an incorporated city, town, township, borough, or village or an unincorporated area of a county or parish."

The NFIP seems to focus on a community as a geographic entity that has powers (i.e., land use zoning authority [FEMA, 2005a]), at a minimum, to regulate behaviors of homeowners and other real estate interests to take actions that would reduce flood risk. It also may have the power to undertake projects that have the potential to limit flood risk.

As FEMA contemplates the future prospects of CBFI, a clear and explicit definition of the requirements placed on an entity that purchases such insurance would be most helpful. Although the previous definitions of a community are useful, it may be necessary to broaden the current FEMA definition. A town or city would clearly qualify. However, it is unclear whether a geographic area in a city, a gated community, or a business district would qualify. Minimum requirements may be the authority to enter into contractual arrangements and the ability to pay for insurance by assessment or cost levy. Another possible requirement may be that purchase at the entity level results in cost savings.

Rationale for Community-Based Flood Insurance

The implicit arguments for CBFI relate to at least two identified concerns: information and incentives. First, individuals and communities may not be sufficiently informed about levels of flood risk, and more importantly, efforts that would reduce flood risk. Second, even when they are well informed, individuals and communities may not have sufficient incentives to reduce flood risk, stemming from a variety of factors, including the expectation of disaster relief if insurance is not purchased. If insurance is purchased, then its price may not sufficiently recognize the risks associated with the covered entities. If none of these problems were present, then it would be appropriate to have individual property owners purchase insurance on a market, whether from private insurers or the government, just as they purchase toasters, clothes, and automobiles. Flood insurance would be no different.

One justification for devolving additional insurance purchase responsibilities to communities, either down from FEMA or up from the insured themselves, would be that neither communities nor individuals at present properly recognize some costs in the system. Perhaps relevant information is not available to the parties that must take actions to control flood risks, or perhaps incentives are insufficient. A second justification would be that placing insurance through communities would reduce or eliminate failures in providing and recognizing relevant information on flood risk. To take more appropriate actions the academic field and literature on "agency theory" can provide some guidance.

Agency theory looks at a broad array of situations when one individual (an agent) acts on behalf of another (a principal). The community, or more appropriately the decision makers in a community, are agents for the stakeholders in the community (residents and other asset holders), who are the principals. The agents should faithfully represent the principals' interests. If they do not, then this represents a situation of failed agency. The attractiveness of community insurance will always be based, in part, on the proposition that principals' interests are faithfully represented by agents. Failed agency complicates enormously the relative attractiveness of community insurance (Pratt and Zeckhauser, 1985).

As an example of agency theory application, consider a hypothetical community, Riverville, whose losses from failing to implement appropriate policies take three forms: (1) expected losses that exceed the costs of their prevention, (2) reductions in insurance charges that are not being realized, and (3) expenditures to reduce risks that fail a cost/benefit test. The argument for getting Riverville to pay for insurance assumes that, for example, given current arrangements, its leaders would not vote to build a levee for $1 million that would save its residents an expected $1.1 million. However, if the community was directly paying a premium that would be reduced if it built the levee, then it would build the levee. Moreover, this would also be true if it paid the premium by collecting premium monies from its residents.

The Responsibility for Insurance is Irrelevant (RII) Proposition

The Coase Theorem from the discipline of economics addresses the common situation where there are two disparate parties—one imposes a negative externality on the other, say a polluter upstream on a river and a laundry downstream (Coase, 1960). The surprising result of the Theorem, is that it does not matter which party has property rights. Thus, if the laundry has rights, then it will charge the polluter for emissions until an additional $1 in cost to reduce emissions just avoids an additional $1 of damages. If the polluter has rights, then it will demand payment from the laundry to clean up to precisely this same point. A flood example might include damming of a river that negatively impacts downstream flooding.

The analogue of the Coase Theorem in the flood insurance context, where the interests of the two parties (community and residents) would be expected to coincide, is that which party bears responsibility for insurance does not matter. Call it the RII Proposition as an acronym for "Responsibility for Insurance is Irrelevant." If the RII Proposition is satisfied, community insurance would neither help nor hurt. (The technical features of the RII Proposition are described in Appendix B. Appendix B identifies the parties to the insurance decision, and then explains how appropriate incentives can lead their decisions to produce an efficient outcome in the flood insurance case.) To understand why insurance arrangements might fail, it is important to first recognize the properties of and conditions for an efficient outcome.

Appendix B shows that if both the community and residents must take actions to reduce the risks of and damages from floods, then an efficient outcome can result given reasonable behavior. Each party, that is each resident and the community, must spend until an additional $1 of expenditure just reduces expected losses by $1. To produce such an outcome, two properties must hold: (1) individuals must take the actions they would if they bore full responsibility for their own losses, and (2) the community, in choosing its actions (expenditures), must take full account of the losses to its residents.

Appropriate actions from the individuals (residents) can be achieved in two ways. First, they could be required to pay any incremental costs that their own actions impose on the costs of insuring their losses, for example, if they choose to live in a high-risk area. Second, the community could impose regulations that appropriately control residents' actions, for example, by imposing building codes that efficiently control the risk and magnitude of flood losses to each resident.[1]

For these conditions to hold, of course, the residents and the community must also be well informed about how their exposure to flood risks and losses is affected by the actions that they take. If the community employs the regulatory approach to secure appropriate actions from residents, then it must understand the costs to residents in adhering to such regulations, for example, how much they would pay to live in such areas.

Why the RII Proposition Might Fail to Hold

As an example, one could assume that given the current system where insurance is purchased at the individual level to an inefficient outcome in Riverville. Such a failure would refute the RII Proposition. The RII Proposition starts by observing that the community has no independent interests of its own, and is expected to act as a faithful agent and take full account of its residents' interests, implying that Riverville and its residents would have fully congruent interests. In circumstances where the RII proposition may fail to hold, CBFI may be superior or it may be inferior depending upon the circumstances.

Why might the RII Proposition fail to hold? This is a more general and much more important question about public finance and public choice. Why might governmental entities not appropriately represent the interests

[1] Prohibiting residents from living in high-risk areas might be inefficient—and would be, if the residents paid substantial amounts to live there. Further, FEMA has no legal authority to do this.

of its residents who were paying their insurance bills themselves,[2] or why might those residents not make optimal insurance purchases of insurance? And more broadly, would performance be improved or diminished if insurance were instead purchased at the community level?

Any argument for CBFI must start with an argument that the RII Proposition will fail to hold in practice. Failure of the RII proposition, however, is not sufficient to justify CBFI. It is required as well that insuring at the community level will work better than insuring at the individual level. Below are eight reasons why the RII proposition might fail to hold:

1. **Free Riding.** Residents might not purchase flood insurance because they expect disaster relief to bail them out; hence they would take a "free ride" and get coverage for free. To make CBFI preferred, the community would have to be better able than the current program to require its residents to buy insurance, to avoid living in certain areas, or to build to certain standards in flood-prone areas.
2. **Alternative Financing.** The community might choose a different mode of paying for the insurance than charging residents on the basis of their actuarial costs. For example, it could just use the general tax base. Such a shifting of responsibility might secure funding in a way that some would consider more equitable, but it would sacrifice efficiency, because residents would no longer have appropriate incentives to take their own damage-avoiding measures. And some, including many residents, would consider such a cross-subsidy program less equitable, quite apart from its efficiency implications.
3. **Externalities.** If owners or renters in the community would prefer that others purchase flood insurance, then such purchases would convey a positive externality. Such an externality could flow from altruistic or paternalistic feelings, or from real losses—for example, deterioration of the community—if the assets of others are lost and not replaced. Self-interested owners would not take account of this externality when deciding whether to purchase insurance. A community that accounts for all of the impacts on its constituents would account for all externalities.
4. **Behavioral Mistakes by Residents or the Community.** The residents may fail to purchase insurance because they underestimate flood

[2] In economics, the agency problem is where there is a need for the community to take optimal actions on behalf of its residents. In this situation the community is the agent and the residents are the principals. Other prominent agency problems arise between a lawyer agent and client principal, or worker agent and owner principal. The principle explanation for losses from efficiency in those cases is that the agent and principal have divergent preferences. That explanation would not apply here, at least not if the community considered its interests to be the same as its residents' interests.

risks for example because there has not been a recent flood. The whole panoply of heuristics and biases discovered in the past few decades by psychologists and economists, namely human tendencies that lead them to make decision-making errors, could lead to such mistakes (Tversky and Kahneman, 1974; Kahneman, 2011). So too, community officials might fall prey to these mistakes.

5. **Community Pricing Is More Appropriate.** Current pricing for flood insurance deviates from risk-based pricing. For example, it has grandfathering exceptions.[3] It is also possible that the community, being closer to the problem and also utilizing federal information, can assess risk more accurately than can a distant federal agency. When assessing at the community level, to establish a community premium, the community decision makers need not be concerned if some residences are over assessed and others under assessed. Determining a reasonable overall level is important.
6. **Political Distortion.** Community political figures, such as mayors and city councilors, are elected with support from part of the community. Often that support comes differentially from different locations within the community. In making flood-control decisions, they may cater disproportionately to their supporters, implying that such decisions would be inefficient overall.
7. **Out-of-District Owners Undervalued.** In a special important case of reason 5, political figures might well feel that external (out-of-district) owners[4] deserve lesser consideration than residents. Thus, when deciding what protective measures to undertake, they might give inadequate weight to such external owners' damages or insurance payments. Community insurance, in theory, would account for damages to all dwellings in the community and would make underweighting external owners less likely.
8. **Greater Community Attention to Expenditures That Are Budgeted.** The efficiency condition (see Appendix B) requires that the community value \$1 to a resident equal to \$1 of its own expenditure. It may not do so, for a variety of reasons. First, it may not notice or may under notice those dollars. Few mayors or city councilors may know what their residents are spending for flood insurance, for example. Second, officials are concerned with what the voters can easily monitor. If the town spends a great deal to rebuild a marsh, then that will be noted, perhaps much more than will insurance

[3] Such exceptions arise when, for example residents have mistakenly not been charged for something in the past, they are also not charged for it in the future.

[4] This report uses of the term residents elsewhere in this section includes out-of-district owners.

savings to residents. Third, the benefits of community spending may be highly unevenly spread. If only a small minority of individuals is at risk of flood, then the town may be hesitant to make an expenditure that would be shared by all residents, but would benefit only a minority. Fourth, there are many other reasons why the community might spend differently if the insurance premium is paid out of its budget, as opposed to its residents' budgets.

These eight reasons, and some subcategories within the reasons, explain why the RII Proposition may not prevail. Other observers would add further reasons. Some of the above reasons, by themselves, would indicate that insurance at the community level is superior, others that it is inferior, and others that it depends on the underlying empirical situation. The most important lesson to emerge from this discussion is that any analysis of CBFI should begin with understanding of the RII Proposition and whether the evidence points in favor of, or against, insurance at the community level.

Spillovers Across Communities

Regardless of the flood insurance arrangements that are in place, a community has a responsibility to take action so that the sum of expected flood damages and/or insurance premiums avoided for the residents just equals, or exceeds, the cost of taking additional actions. This condition is appropriate if the community's actions have no impact on other communities and their residents. However, many actions taken by community A may have a positive or negative consequence for nearby community B. For example, the construction of levees in upstream locations can raise flood stages downstream, and thereby contribute to increased downstream flood damages. The spillover could be positive, by contrast, for community D if nearby or upstream community C restored a wetland.

The formulation above assumes that each community optimized for itself. Moreover, this assumption would be satisfied whether insurance premiums were paid at the community or the resident level. With self-interested decisions by communities, which are to be expected, these spillovers will either be ignored or taken into account insufficiently. One potential solution would be to make a larger unit, a collection of communities that mainly impact each other, the unit of decision. Another potential solution would be to have the community receiving the spillover to pay (if positive) or get paid (if negative) so the spillover would be "internalized," that is taken into account by the community creating the spillover. This latter system would work if the aforementioned Coase Theorem applied when communities were the players, which seems like a heroic assumption in this context.

How CBFI Might Provide Solutions

A CBFI option, like any other policy option, has both strengths and weaknesses. Such an option might provide solutions—but not the sole solution—to recurring issues within the NFIP. The potential solutions identified and discussed in the following sections are whether CBFI can reduce administrative and transaction costs; increase takeup rates; and promote mitigation and floodplain management.

Reduce Administrative and Transaction Costs

Any insurance program entails administrative and transactions costs. Rates must be set, premiums collected, delinquent accounts pursued, and claims settled. Other NFIP costs include the federal policy fee and surcharges. Costs can vary depending upon the size of the institution that administers the insurance and the extent to which specialization may achieve savings. Many small employers contract with specialty firms to provide insurance-related programs because such outside firms may have lower administrative and transactions costs. In some circumstances, CBFI may entail administrative and transactions costs that are considerably lower than the aggregate of such costs to individuals who purchase insurance independently. In these instances, CBFI could be a less expensive option.

The extent to which a CBFI option has the potential to reduce transaction costs will depend upon the overall design of the option. Thus, for example, the NFIP currently pays 15 percent to write-your-own (WYO) agents, which reflects the cost of writing individual policies and provides agents an incentive to promote the policy option. A single policy covering an entire community may entail much lower costs because of economies of scale and the absence of the need to create incentives. Some costs will remain, however, including the costs of collecting data, setting rates, and settling claims.

The extent to which CBFI would reduce claim adjustment and processing costs is unclear. Adjusters presumably still need to evaluate damage and estimate repair costs structure by structure. There would still need to be some type of appeals process for policyholders who are dissatisfied with the initial insurance award. The administrative costs required to design and set up the program would also need to be considered. The return on this investment may not be large if few communities decide to enroll.

Increase Takeup Rates

One frequently expressed concern about flood insurance is that many high-risk properties go uninsured: takeup rates are low. Low takeup rates can occur for a variety of reasons:

- Owners without mortgages are not required to purchase insurance.
- Owners with mortgages, for whom insurance is mandatory, still might not purchase.
- Although owners or renters should buy insurance, they mistakenly do not.
- Mispricing makes it undesirable for some owners or renters to buy insurance.
- Owners or renters avoid insurance because they expect government bailouts should there be a flood.

Of course, some properties with very low risk—a house on a hill, a business far from any water—may not need to be insured. These are properties for which the administrative costs of issuing a policy outweigh any risk-spreading benefits.

This problem could be readily dealt with at the community level, assuming that the community has the power to require its constituents to purchase coverage. The simplest way, of course, would be to have the community just buy a policy that covers all properties (or all properties that merit insurance). This could be done even with minimal requirements on what constitutes a community, say a gated community or business district. If CBFI were a catalyst for increased takeup, then it could enhance resilience (when paired with floodplain management and mitigation activities) to flood events and, depending on how premiums were set, provide incentives to mitigate more properties.

The mandatory requirement to purchase flood insurance increased takeup rates, but compliance is not complete and takeup remains low among those not subject to the requirement. The issue of mandatory purchase is both complicated and controversial, especially in a world where mortgages are often securitized and resold. Ongoing attention has been paid to enforcing the mandatory purchase of flood insurance by homeowners (within the SFHA) that have a federally insured mortgage. Efforts to increase enforcement by focusing on the lending sector have been partially successfully (NRC, 2015). For example, the NFIP increased the responsibilities of the lenders by requiring them to notify property owners when their policies lapsed. The National Flood Insurance Reform Act of 1994 also established penalties for lenders who did not carry out specific requirements (NRC, 2013). A CBFI option may help with such an issue.

Promote Mitigation and Floodplain Management

One of the primary arguments in support of CBFI is that communities will take more actions to mitigate risks if they themselves must purchase insurance. This could be accomplished in two ways:

1. The community could undertake actions on its own, such as restoring a marsh that can mitigate storm flows and thereby reduce flood risks downstream.
2. The community could require owners to take actions that reduce risks, for example by not locating in a flood-prone area.

However, this potential for action is not a sufficient argument for CBFI. It must also be the case that the community would not take this action if insurance was purchased given current arrangements. The reason why it might and might not take such actions was the subject of the rationale for CBFI above.

Floodplain management and flood hazard mitigation are long-term activities, which may not garner political backing during the short tenures of elected officials. Local communities receive the property tax benefits from development but do not pay insurance premiums for structures developed in flood hazard areas. Often, local officials make land use decisions based on short-term economic gain that in the long term increases flood hazard exposure. Public support for flood mitigation peaks and fades with flood events and tends to be sustained only after a community has repeated flooding. A CBFI option has the potential for creating a monetary measure to enhance community floodplain management and mitigation action over the long term without waiting for flooding to become chronic. Flood mitigation has increasingly become the responsibility of local decision makers (Brody et al., 2010). Community-level interventions and particularly nonstructural mitigation can reduce flood losses. For example, there is empirical evidence to suggest the importance of strong organizational capacity—at the local level—to implement flood mitigation strategies (Brody et al., 2010). Therefore, it may be helpful for a community-level option to consider the contextual conditions and capabilities of local jurisdictions (Brody et al., 2009).

CBFI also has the potential for making local officials more aware of their land use decisions. The structure of payment for insurance by citizens could be devised to encourage them to take action to reduce their risk. Some local government units, such as drainage and levee districts, are organized to provide flood protection. The residual flood risk in areas protected by levees is acknowledged in NFIP reform legislation and the U.S. Code pertaining to the National Flood Mapping Program. A CBFI policy made available to local units of government could provide coverage for residual risk. A strategy suggested for increasing private-sector involvement included the option of the NFIP providing residual insurance (GAO, 2014c). Within this strategy, GAO suggested that the federal government could encourage private sector involvement by providing coverage for the highest-risk properties that the private sector will not insure. Providing residual insurance could increase the program's exposure, but NFIP would be insuring fewer properties.

Key Challenges for a CBFI Option

It is easy to envision a community that purchases insurance optimally on behalf of owners, collects the required monies inexpensively either through premiums or taxes or some combination, undertakes efficient mitigation actions on its own, and requires its constituents to take efficient actions. This, however, may be unrealistic. Taking responsibility for a program of insurance and mitigation is a significant responsibility, and many communities would not be up to the task. The following section identifies and discusses some reasons why CBFI may not be successful, which are lack of community interest, limited capabilities to implement CBFI, variations in the size of the population and geography involved, ability to regulate land use and authority to collect revenue, and goals not achieved, despite adopting CBFI.

Lack of Community Interest

It is unclear how many communities would participate if FEMA put a CBFI option into practice. From a practical perspective, community leaders might be reluctant to participate in a CBFI option, because it would require adjustments from the status quo. Even if CBFI were financially and economically attractive, there would be some losers. For example, if the community simply imposed the required premiums on owners, then those who had considered it economically desirable to go uninsured (e.g., free riders) would bear costs that they had previously escaped. If some or all of the premiums were paid through taxes, then individuals whose tax burdens increased above the level of their current insurance costs would likely object. And if the premiums were paid simply from tax revenues, then leaders would be confronted with constituent complaints, which characteristically occurs when taxes are increased. In addition, current leaders might be reluctant to purchase for others: the vast majority of the time it does not pay off. Thus, the community would incur a cost that would have little chance of being paid off during the leaders' time in office. Political leaders tend to operate with very short time scales. Thus, the prospect of requiring people to purchase insurance against an event that might occur in the distant future might be unattractive. The same argument would apply to undertaking mitigation measures. However, if a marsh is restored or a river bank raised, then at least the constituents can see what they paid for. The promise of protection is visible, and retains its value even if there is no flood.

Capabilities to Implement CBFI

Even in the absence of political concerns about undertaking CBFI, community leaders may have to assume responsibilities to successfully implement CBFI. Some of these responsibilities may require capabilities that many communities do not currently possess. Consider, for example, the simple case where the community pays the premiums but passes the charges to individual property owners. The community would require expertise in pricing that involves risk assessment and property valuation. It would also require the capability to oversee compliance with mandatory purchase requirements. Responsibility for mandatory purchase and maintenance of flood insurance could remain the responsibility of the lender. Additionally, the community would need to perform more traditional roles of establishing and enforcing building and zoning standards for areas susceptible to floods. These latter tasks would require engineering expertise and processes for managing and resolving appeals.

To help address issues related to community capability to carry out a CBFI option, multi-jurisdictional programs based on experience with hazard mitigation planning under the Disaster Mitigation Act of 2000 (DMA) (Box 4-1) were considered. Communities (e.g., counties, municipalities) can chose to consolidate or create separate mitigation plans. Communities sometimes cover both county and municipal jurisdictions, often called multi-jurisdictional plans under DMA. Small municipalities, for example, with limited capability to prepare and implement plans often choose to combine their planning with counties. These joint actions may help in coordinating spillover effects of different jurisdictional actions, setting insurance rates, and reducing transaction costs.

Variation in the Size and Homogeneity of the Populations Served

Some NFIP-participating communities have many policies-in-force, while others have few policies-in-force; therefore, there is little evidence for an average NFIP community (see section on "Participating Communities," Chapter 3). This set of observations suggests that small communities might not have the executive/professional capability to oversee a CBFI option, or if they did, it would be at too small a scale to be efficient. If a community turned program management over to a third party, then it would just substitute that party for the federal government. However, this does not mean

BOX 4-1
Disaster Mitigation Act of 2000

Section 203 establishes a "National Pre-Disaster Mitigation Fund" in order to carry out a program that will

- Provide technical and financial assistance to States and local governments to assist in the implementation of pre-disaster hazard mitigation measures that are cost-effective and designed to reduce injuries, loss of life, and damage and destruction of property, including damage to critical services and facilities under the jurisdiction of the States or local governments.

Section 322 provides a new and revitalized approach to mitigation planning by specifically doing the following:

- Establishes a new requirement for local and tribal mitigation plans.
- Authorizes up to 7 percent of the Hazard Mitigation Grant Program (HMGP) funds available to a state to be used for development of state, local and tribal mitigation plans.
- Provides for states to receive an increased percentage of HMGP funds (from 15 percent to 20 percent) if, at the time of the declaration of a major disaster, they have in effect an approved State Mitigation Plan that meets the factors in the law.

SOURCE: Illinois Emergency Management Agency, 2014.

that large communities would be ideal for implementing CBFI. Such communities might do well in passing through insurance premiums, but there could be difficulties when cross subsidies are involved (e.g., from high-lying to low-lying areas). If the cross subsidies are between different ethnic or income groups, then the difficulties could be greater still. In a small community, the residents of the two different areas might have more connection to each other or have more concern for each other than they would in a large metropolis that comprises a city, a single political unit. Indeed, it might be optimal to even have protective expenditures financed on a smaller scale than a large city. For example, the lack of concern of most residents for those living in a high-threat area might prevent the undertaking of efficient protective expenditures. (Of course, that phenomenon is already at work, given that such expenditures would come out of the budget of the city as a whole).

In short, given the administrative capabilities required for running an insurance program, communities that are too small would likely be ill equipped. However, as communities grow they tend to become more diverse in their risk levels. In addition, there tends to be more social variability and a consequent wider range of preferences in such situation than in smaller communities. This suggests that very large communities might also have difficulties administering insurance, given political divisions. Should the NFIP offer a CBFI option, the optimum scale may be midsized communities, which may be better suited to balance needs for administrative capability against the dangers of insufficient concerns of some citizens for the well-being of others (Tiebout, 1956).[5]

Concerns about aggregate risks, and the need for a large population to spread them, are not part of this consideration. CBFI would be paid out of a federal pool, and millions of insureds in hundreds of communities would be covered. Even in hypothetical Riverville, with its two residents, aggregate risk would not be a concern; its risks would be laid off at the higher federal level. No supporter of CBFI has been proposing that communities also self-insure. Were they to do so, aggregate risk would certainly be an overwhelming concern.

Land Use and Revenue Collection Authority

The discussion above implicitly and explicitly assumes that communities would be able to regulate land use to achieve flood protection at the lowest possible cost, accounting for all cost elements. Absent this authority, individuals might choose to live where risks are high, particularly if they could expect—perhaps incorrectly—either subsidized insurance rates or post-flood disaster relief. If the community lacks the authority to restrict land use, whether insurance is at the individual or community level, then considerable inefficiencies will result.

A second implicit assumption is that communities contemplating CBFI would have the authority to raise revenues to cover the premium. If the community were merely a pass through, then it would collect insurance premiums from its residents. If the premium were to be financed in part by taxes, then the community would require the ability to raise taxes to cover the expense. Collecting premiums could be challenging administratively. Raising taxes, particularly if the beneficiaries were a relatively small group, might be difficult politically.

---

[5] The Tiebout model suggests that forces sort individuals so that those within a community will have fairly similar preferences. Unfortunately, geography often implies that different locations within the same community will be at very different risk levels. Thus, even after sorting, although residents may be similar in terms of preferences on taxes and school spending, they may have quite different preferences on whether a wetland should be created or flood insurance subsidized by the municipality.

Political Distortions

As mentioned above, political processes can never fully reflect the views and preferences of constituents. The problem stems in part because, politics aside, methods for aggregating individual preferences fail to capture relevant information, for example how strongly they care about an issue. Even if all preference information were known, however, the electoral process assures that elected officials will feel more beholden to some people than to others; hence the former will have their values over-weighted. There is considerable evidence in addition that those who participate in particular decisions are those who care the most. In effect, the intense interests of a minority may outweigh the diffuse interests of the majority. The exact balance of these forces in any situation is unclear, and it may lead the high risk to exploit the low risk, the rich the poor, or vice versa in both cases. What is clear is that politics can never assure an efficient outcome. The critical question in this context is how well a community-based process, however designed, performs relative to the current imperfect situation.

Additional Challenges

The implementation of CBFI as a policy option presents additional challenges. First, many communities that should adopt it will not do so, perhaps because leaders are influenced by politics or do not want to assume new responsibilities and tasks. Second, some communities that lack adequate capacity and resources to effectively implement and administer CBFI may nevertheless choose to do so. They may overestimate their capabilities to administer the program, or their ability to build those capabilities swiftly. The result would be a poorly administered program in these communities. Possibly, there would be swift reversion to the current program. If CBFI were implemented, then there may be potential for delay in disbursements to individual property owners or renters. An additional administrative layer may extend this period even longer, particularly for those who are un- or underinsured, which may lengthen the recovery period and make it more variable for the most socially vulnerable households and neighborhoods within a community.

Risk and Uncertainty

Insurance companies, and other insuring entities, deal with risk in a variety of ways. The first line of defense is to properly underwrite risk, which begins with estimating the probable frequency and severity of losses associated with risk exposure. Generally, a new risk exposure is added to an existing pool of exposures if it meets the insurer's underwriting criteria and does not cause the insurer's overall risk to exceed its appetite for risk. Pooling a larger number of independent exposures enables the insurer to more accurately estimate total losses to be paid and to establish appropriate premiums for coverage. Diversifying across geographic areas or lines of business can also help reduce risk, if the areas or lines of business are independent. An insurer's risk can be defined by a distribution of expected losses which, more fundamentally, captures unknown outcomes whose likelihood of occurrence can be measured or calculated.

In addition to bearing calculable, objective risk, insuring entities face uncertainty in the form of events that cannot or at least have not yet been described. Various actions can be taken to address uncertainty. Through time, of course, uncertainty regarding some types of losses can be reduced by learning more about them, that is, collecting and analyzing event frequencies and severities, and better understanding correlations between losses. Various manifestations of climate change, described in Box 4-2, are examples of uncertainties that could be faced by flood insurers. These are detailed in general in the following paragraphs. Specific examples drawn from the climate change example are identified and detailed in Box 4-2.

To deal with uncertainty, insurers may contractually limit coverage to known events (e.g., through a named perils policy) or limit the amount of losses to be paid under the contract. The NFIP currently employs both of these approaches.

Yet another approach for dealing with both risk and uncertainty is to transfer a portion of risk to a reinsurer. In such an arrangement, an insurer transfers its catastrophic risks to a reinsurer or reinsurance facility in exchange for a known premium amount. Because the insurer is primarily concerned that losses might exceed a target level for maintaining financial solvency, a reinsurance contract could be purchased to cover losses exceeding this target. In addition to transferring the known risk, such a contract would, consequently, also transfer to the reinsurer the uncertainty associated with unknown catastrophic events.

The methods of managing risk and uncertainty are more problematic in the context of floods because of the fundamental nature of the risk. Diversification across geographic areas can help to reduce risk, but pooling a large number of exposures that are correlated does not help the insurer reduce its risk because additional exposures do not reduce the variation of expected losses. This largely explains why individual private insurers do not write flood risk exposures: their ability to diversify geographically is often limited. Uncertainty, in the flood risk context, stems primarily from the unknown nature of future flood events (e.g., changes in event frequency and severity due to climate change), but also from uncertainty about the effectiveness of mitigation activities and neglected contributors to damages such as failing structures or storm drainage systems.

BOX 4-2
Risk, Stationarity, and Climate Change

Estimating flood risk and assessing the expected annual losses from flood damages at any location is difficult. In a classic paper focused on the potential of flood insurance, Langbein (1953) noted that the existing record of flood discharges for a river is only one sample of the entire history of that river. Flood characteristics differ in differing periods, which means that damage assessments based on a specific period of record will be biased depending upon the level of flood activity in the period in question. This bias makes the estimation of flood damages for insurance purposes problematic. One approach is to develop confidence bands that will give a reasonably safe estimate of flood damages. However, if the confidence bands are broad, then the insurance premiums may be very expensive. If they are unduly narrow, then premiums may be insufficient to cover payout and related costs and the insurer may suffer losses. The fundamental challenge is to find an economic balance between the two. The problem is further complicated by non-stationarity due to climatic, as well as landscape-based trends.

Over the past century or so, flood protection and management measures have been based on the proposition that past climate provides a reasonable depiction of what we might expect now and in the future. Thus, in the past, the probability of floods of different magnitudes has been assumed invariant or stationary. That is, it can be estimated from the historical record and the uncertainty of those estimates can be determined from statistical theory. Today, it is well understood that this assumption of climatic stationarity can no longer be defended scientifically. With floods and other climatic phenomena, past is not prologue. Moreover, there is no substitute assumption that would reliably define the nature and rates of change in risk levels.

Several factors explain why assumptions of hydro-climate stationarity are no longer scientifically defensible. Changes in any given watershed, such as urbanization and land drainage, and changes in river channel, such as dams and channelization, alter the way in which flood waters are stored and transmitted downstream. There are known methods for assessing the impact of these factors on flood risk. These analytical tools are expensive to employ, but community-

In CBFI, the task of managing risk and uncertainty would fall on the risk-bearing entity, the federal government, as it does in the current NFIP. Moving to a CBFI option, however, might yield new opportunities for managing risk and uncertainty. For example, the NFIP could require that communities seeking coverage provide a comprehensive analysis of the flood risk.

The above section discussed topic areas and questions that require further evaluation when assessing the strengths and the weaknesses of a CBFI

based flood insurance (CBFI) covering a large number of properties may make them worthwhile. Climate change, about which the science is much less certain, is the other major factor driving non-stationarity. On the one hand, a warmer atmosphere can hold more water and this could lead to even larger storms that have increased intensities of rainfall stormwater runoff. On the other hand, a warmer climate means more annual precipitation falling as rain instead of snow; when snow does fall, it will melt more quickly leaving less water to create floods in the spring when temperatures rise and rain-on snow events occur. A warmer climate can lead to drier soil conditions, increasing the amount of infiltration from a given storm and hence less runoff. Climate change, then, could either increase or decrease flood risk depending upon local conditions and the hydrology of local floods.

The observational record of large storms in the United States is not consistent with the predictions of climate models. The lack of scientific certainty does not mean that climate change will not influence flood risk or flood damage. Rather, it means that science is currently unable to provide reliable forecasts of the types and direction of change that may occur on any river or in any given community (Milly et al., 2008; Peterson et al., 2013). The Intergovernmental Panel on Climate Change (IPCC) states: "...there continues to be a lack of evidence and low confidence regarding the sign of trend and magnitude and/or frequency of floods at a global scale" (Stocker et al., 2013). Recent studies have shown increases over the past several decades in the size and/or frequency of floods in parts of the northeastern and upper Midwest of the United States, but decreases throughout the west and particularly the southwest. Other parts of the United States show no clear pattern of decrease or increase in the magnitude or frequency of floods over the past several decades.

Despite the scientific uncertainty about the impacts of climate change, many actions could be taken to mitigate against flood risks at a given location. For example, the scientific evidence indicates strongly that the observed increases in flood losses are mostly explained by what is done to or on the landscape and this will be true for decades to come (Kundzewicz et al., 2014). CBFI may well encourage citizens to become politically engaged in understanding and mitigating the ever-growing level of national flood losses by focusing on improved landscape management.

option. It also considered aspects of risk and uncertainty that pertain to a new insurance option. The next section discusses design considerations of a CBFI option, if it is pursued.

DESIGN CONSIDERATIONS

Fundamental considerations in the design of CBFI include the following and are discussed in detail below:

- Who bears the cost of risk and how is it shared
- Who writes the policy and adjusts for losses
- Coverage limits, standards, and compliance
- Underwriting, pricing, and allocation of premium costs
- Administrative capabilities
- Confirming compliance within mandatory purchase requirement
- Pricing expertise, including valuation of mitigation measures

Who Bears the Cost of Risk and How Is It Shared?

When feasible, risk of any nature, catastrophic or not, is ideally transferred to the entity that is in the best position to bear it. This is done largely for reasons of solvency, to ensure that the entity has sufficient capital, or access to capital, in the event that larger-than-expected losses are incurred.

The fundamental nature of floods complicates private insurers' ability to underwrite flood risk. Specifically, floods constitute a catastrophic risk which, for noncommercial residential properties in the United States, is typically not insured by private companies except to the extent that they underwrite excess flood coverage beyond the NFIP's policy limits. Adding full coverage of flood risk to their portfolios would present several challenges, because it would substantially increase their overall exposure to catastrophic losses. One such challenge results from the compounding of flood risk associated with tropical storms and hurricanes with the wind damage from these events.[6]

The federal government (including the U.S. Treasury) is the risk-bearing entity for all flood insurance policies sold through the NFIP. Insurers that participate in the WYO program do not bear risk. Individual homeowners bear risk only in the form of deductibles and losses that exceed policy limits.[7] The movement to a community-based option would conceivably shift risk-bearing to communities, private insurers, or individuals depending upon how it is structured. Although, as a federal entity, the NFIP may be best positioned to bear the risk, movement to a CBFI option allows for a reexamination of how some risk might be transferred to and/or shared by other stakeholders. The manner in which flood risk is shared among different parties will affect their respective incentives for mitigating losses.

A community-based option could allow for the transfer of some flood risk to the community (e.g., through a high deductible paid by the community), but consider the community's risk-bearing capacity. A larger com-

[6] Wind insurance is typically underwritten by private insurers or state-run wind pools.

[7] Policyholders with pre-FIRM subsidized rates may purchase a minimum deductible of \$1,500 if the building coverage is less than or equal to \$100,000; if the building coverage exceeds \$100,000, then minimum deductible is \$2,000.

munity, for example, does not necessarily have a greater ability to bear risk. In addition, not every type of community has the ability to raise capital (e.g., a flood district vs. a county). Shifting some risk to the community may incentivize the community to mitigate flood risk. Deductibles and risk-based premiums that reflect mitigation behavior in a timely manner would provide incentives for a community to undertake mitigation.

In terms of the risk borne by individuals, those individuals with flood insurance bear risk in the form of deductibles and policy limits. Deductibles, policy limits, and risk-based premiums provide incentives to mitigate; but many potential mitigation measures may be out of the control of the homeowner (e.g., reinforcing a federal levee). Property owners can undertake some measure to make their homes/buildings more flood resistant (e.g., elevating structures), but these measures can be costly.

Who Writes the Policy and Adjusts for Losses?

The transfer of flood risk from the individual homeowner (or community) to an entity better suited to bear the risk (NFIP and community) requires a contractual agreement which is typically an insurance contract. Fundamental considerations in the design of this contract are the coverage limits, standards for coverage, cost-sharing mechanisms (i.e., deductibles), and the insuring agreement itself, which outlines the coverage provided.

A CBFI option could capitalize on the already-existing expertise and capability of NFIP and WYO insurers for writing policies for individual homeowners. That is, CBFI could involve the NFIP and the WYO insurers in writing policies and collecting premiums, but at the community level rather than the individual level. The processing of applications and claims from communities versus individuals would likely require additional administrative effort. In theory, however, the claims process may not change dramatically. For example, the insured incurs a loss and notifies the policy issuer (i.e., the community), who will notify whoever they got the policy from (either NFIP directly or through a third party), who will contact the adjuster, who will report on damages to the party responsible for sending claim payment to the property owner.

Coverage Limits, Standards, and Compliance

Another design consideration for CBFI is the extent of coverage that will be provided under the community policy. Currently, individuals are provided a limited set of coverage options. Homeowners may elect coverage for building property up to $250,000 and personal property up to $100,000. Within these coverages, they also have a choice of deductible amounts. Under CBFI, new limit guidelines would have to be established for policies that are written for communities, with such limits and choices of deductibles potentially depending on the size of the community, the nature of the risk (e.g., type of flooding), existing infrastructure, and other community characteristics, for example, floodplain district versus a county.

Movement to a CBFI policy option may provide an opportunity to reconsider flood exposures; that is, defining the community's exposures as extending beyond individual homes to public infrastructure (e.g., bridges). To the extent that the community policy can cover a variety of exposures, additional consideration is needed to establish coverage standards and limits that reflect this expanded group of exposures.

If designed to include all individual homes in a defined flood-prone area, then a community-based policy has the potential to achieve full compliance with lenders' mandatory purchase requirements. This is because all properties included in the community policy would have the required coverage by design and, furthermore, the ability of individual homeowners would not be able to opt out of coverage. This constraint would depend on how the community establishes the rules of participation. For example, if the community charges homeowners a monthly premium or collects payments in the form of taxes. These considerations are addressed further below.

Underwriting, Pricing, and Allocation of Premium Costs

How the CBFI is priced will have important implications for the entity that bears the risk, the community, and the individuals in a community. If an insurer were to underwrite and bear all of the risk of a CBFI policy, then it would need to charge an actuarially sound premium that reflects all expected costs, including an appropriate catastrophe loading, which could be used to purchase reinsurance in the private market. If the NFIP were to provide reinsurance to the insurer, then the premiums charged by the insurer would only need to account for whatever the NFIP would charge for this reinsurance. This assumes that the NFIP reinsurance would cover any losses that would be considered catastrophic. There is also the issue of how the premium of a community policy would be allocated among the properties covered under the policy. If a private insurer were to bear the risk assumed under the policy, then its primary concern would be that the premiums would be sufficient to cover all of the costs associated with underwriting the policy. This assumes that insurers would not be exposed to higher than expected costs due to moral hazard or adverse selection. If the NFIP were to underwrite the risk assumed under the policy, then this issue would likely not be a matter of concern to the insurer if its role is confined to servicing the policy.

Other questions arise with respect to how the premium of a community policy would be allocated among the properties covered under the policy. A detailed discussion of alternative cost allocation and financing schemes is beyond the scope of this report, but it is useful to consider two alternative approaches that would represent the opposite ends of a spectrum of cost allocation/financing schemes. With the first approach, the community charges all property owners an amount that depends on property value and flood risk. This could be accomplished through special assessments, a property tax surcharge, or other financing methods. However, this approach would likely be viewed as highly unfair by low-risk property owners and would exacerbate the moral hazard created by the NFIP's current pricing structure. With the second approach, the community allocated the premium in such a way as to require each property owner to pay an amount commensurate with his or her risk of loss. This approach would likely be viewed as fair by low-risk property owners, but it would constitute a departure from the NFIP's current pricing structure and would be opposed by property owners who currently pay discounted/subsidized premiums under that structure.

Ultimately, CBFI requires consideration of how underwriting connects risk-bearing to incentives for mitigation. Pertinent concerns include whether entities and/or individuals are allowed to opt out of participation, or whether their ability to participate is restricted, for example, by requirements for certain mitigation activities. Rules governing participation have important implications for the pricing. The ability of individuals to opt out of a CBFI policy option could result in greater adverse selection.

Administrative Capabilities

If insurance contracts remain the vehicle for transferring risk, then private insurers would remain the most efficient entities for handling administration of contracts. Communities (perhaps regardless of definition) would likely not have adequate expertise to administer policies. However, some definitions of community include entities that could effectively and efficiently collect the revenue needed to pay for a community policy through means such as special assessments, property taxes, and other means.

The majority of NFIP flood insurance policies are written and serviced by WYO private insurers. Typically, several WYO companies may operate in the same community. The bundling of all of the flood risk in a community into a single policy begs the question of who would write and service the policy. Servicing a flood insurance policy after a flood disaster requires knowledgeable professionals to visit sites to assess damage. A single provider may not have the staff capacity to address, in a timely manner, all the needs of a community after a flood event.

Alternatively, community purchase of a base policy (presumably directly from the NFIP) would provide a minimal level of community-wide coverage, lowering the premiums for individual policyholders and leaving in place the current system of involving WYO companies. This design consideration requires no changes to the existing administration system.

Confirming Compliance with Mandatory Purchase Requirement

Under the current NFIP, policies are written and administered and claims adjusted by WYO insurance companies and, to a lesser extent, directly through FEMA. When flood insurance is required because the structure is collateral for a federally backed mortgage (mandatory purchase), the required coverage is for the amount of the outstanding loan (with maximum coverage limits). Lenders have the responsibility to enforce the mandatory purchase of flood insurance for structures in SFHAs; the level of coverage varies with the amount of the mortgage. Additional coverage may be purchased at the discretion of the individual. The level of any non-mandatory purchase of flood insurance, such as replacement cost, renters insurance, and contents insurance is the choice of the individual.

Complying with the mandatory purchase requirement is currently a loan-by-loan, structure-by-structure process. The administrative aspects of monitoring compliance with a CBFI policy option in place may depend on coverage provided through the policy. The mandatory purchase is only for the amount of a federally backed mortgage: a bundled community-based policy providing the minimum required coverage would likely need to maintain some aspect of individual coverage and monitoring, which would be administratively burdensome. Alternatively, if the CBFI option covered all structures to a defined limit, such as replacement costs, then the mandatory purchase requirement would be met. However, using this benchmark for coverage, the aggregate premium for the properties subject to mandatory purchase would likely be higher than the sum of individually required policies for those properties. This leaves unresolved the issue of individuals wanting higher coverage, such as for contents. The important point here is that property owners may be reluctant to pay for CBFI that covers replacement costs on structures. Another, perhaps better, alternative would be a CBFI option that provides a set base coverage amount, for example $30,000. Some homeowners would need to purchase additional coverage to comply with the current mandatory purchase requirement. Individual coverage written by WYO companies would be for mandatory or desired coverage above the $30,000. The administration burden for WYO companies would be comparable to that for tracking CRS-participating communities where premiums are discounted.

Pricing Expertise, Including Valuation of Mitigation Measures

A community might allocate the cost of CBFI in multiple ways. Each method presents unique considerations for pricing and valuing individual mitigation efforts. If private insurers were to underwrite the risk associated with community-based policies, then they would price them on an actuarially sound basis so that premiums would be sufficient to cover the cost of the risk being underwritten and so that incentives for loss mitigation would exist at the individual and community levels. To perform this function, private insurers would have to possess or acquire the information and expertise to price community-based policies to reflect the risk underwritten and the expected savings from mitigation measures. If the NFIP were to assume the risk of CBFI, then presumably it would also price policies to account for the expected savings from mitigation measures. FEMA has the expertise in setting premiums based on flood risk and clearly would have to work with communities to communicate the costs of bundling individual properties into a community policy.

There is a lack of data about which policies are subsidized and grandfathered. A policy cannot be both, although a policy can be eligible for both types (a prospective policyholder has to choose one). One could advance the generalization that structures built before the initial identification date of the flood hazard would tend to be grouped geographically. Similarly, structures built in compliance with the floodplain mapping effective at the time of construction and at a later time identified as being in a higher risk area would also tend to be grouped geographically. Specific conditions must be met for structures to retain a low-risk policy pricing.[8] Given this generalization that structures with grandfathered or subsidized rates would be geographically clustered, high-risk areas with structures dating to a time when flood risk was unknown or considered low could be addressed at the community level for mitigation action consistent with community values. The protection of areas in communities with historic or cultural value is the preferred course of action. At the other extreme, large-scale buyout of structures in other areas would be advantageous if desired by the current residents. Sorting out which structures have subsidized or grandfathered rates might be best undertaken at the community level.

The mechanism used to distribute the cost of flood insurance for the community could be structured to encourage citizens to reduce exposure through individual actions. Stormwater management, for example, offers some interesting examples of community efforts to reduce damages by raising awareness of individual contributors. Some communities have stormwa-

[8] A policy must be in place prior to the effective change of risk zone and the policy must be continued to maintain the lower risk zone rating.

ter management fees, assessed on the basis of the impervious surface area on the property. Many communities work to reduce stormwater damages by offering programs such as vouchers for backflow valves and rain barrels. These strategies could be applied to flood insurance assessments.

SUMMARY

CBFI may help to address some of the current challenges within the NFIP. A basic premise of CBFI is that it may be less expensive and more effective to write insurance policies at the community level. There are many definitions of what constitutes a "community." The NFIP seems to view a community as a political entity that has land use authority. An explicit definition of the requirements placed on an entity that purchases a community-level policy would be helpful.

Understanding of the rationale for CBFI requires some understanding of agency theory, specifically the Coase Theorem from the discipline of economics. The Coase Theorem holds that where parties (both individuals and groups) account for all costs and benefits, markets are functional, information is freely and widely available, transactions costs are zero, and economically efficient outcomes are reached irrespective of with whom the property rights are vested. If the collective economic interests of communities and residents fully coincide and are fully accounted for, then the outcomes of a flood insurance purchase decision do not rely upon which party—communities or residents—bears responsibility for insurance.

In considering practical applications of the Coase Theorem to flood insurance, at least eight reasons explain a Proposition of Responsibility of Insurance is Irrelevant may fail to hold. Two reasons are *free riding*—when some or all residents do not buy insurance because they expect disaster relief to provide adequate post-flood aid—and *externalities*—when self-interested parties fail to account for all of the impacts, most of which are not monetized, of buying or not buying insurance. These reasons help to guide the identification of circumstances when CBFI may be superior, inferior, or either one depending on underlying empirical circumstances. Choosing the CBFI option requires confidence that insuring at the community level will work better than at the individual level.

There are circumstances where CBFI may provide partial solutions to NFIP challenges. Solutions to challenges include reducing administrative and transaction costs, increasing takeup rates, and promoting flood mitigation and floodplain management. Even if CBFI does not effectively address these challenges, it could help in certain areas. For example, moving insurance to the community level would likely enhance attention to risk-reduction activities at that level. Under certain circumstances, however, CBFI may not help to solve NFIP challenges, such as when a community is not interested or lacks the capability to implement CBFI. Successful implementation would likely require communities to restrict land use, adopt complementary flood risk management measures, and raise additional revenue to pay CBFI premiums.

Central to the concept of insurance is protection from losses incurred from uncertain events such as fire, automobile accidents, and floods. Regarding floods and flood insurance, past patterns of climate and hydrology are limited predictors of future patterns. In addition, changes in land use and population sizes influence flood risk and flood damages in uncertain ways. Scientific evidence shows that flood losses are mostly explained by what is or is not done to the landscape; therefore, efforts to improve landscape management are important. For insurance purposes, uncertainties would need to be evaluated on a case-by-case basis.

The chapter also identified design considerations that require further assessment when planning for and designing a CBFI option. These were as follows:

- **Risk Bearing and Sharing**
  A CBFI option could conceivably shift risk-bearing to communities, private insurers, or individuals depending on how it is structured. Although as a federal entity, the NFIP may be well positioned to bear the risk, movement to a community-based option allows for reexamination of how some risk might be transferred to and/or shared by other stakeholders.
- **Responsibilities for Writing Policies and Loss Adjustments**
  Write-your-own insurance agents write policies and collect premiums under the NFIP, but CBFI policies would have to be written at the community level. A range of administrative duties would need to expand to process applications from communities.
- **Coverage Limits, Standards, and Compliance**
  Under CBFI, deductible choices would depend on the community's size, the nature of the risk (e.g., type of flooding), existing infrastructure, and other community characteristics. A CBFI option may provide an opportunity to reconsider flood exposure.
- **Underwriting, Pricing, and Allocation of Premium Costs**
  Several complex issues fall under this topic: the extent of actuarial principles to be used in setting premiums (NFIP premiums are legislatively and administratively constrained; see NRC, 2015); the extent to which catastrophic losses would be reflected in premiums for a given community; and the allocation of premium costs among property owners (and renters) in a given community. The third issue could involve deriving some portion of funding from owners of properties that are not in areas subject to flooding.

- **Administrative Capabilities**
If insurance contracts remain the vehicle for transferring risk, the private insurers would likely remain as efficient entities for handling their administration. Communities (probably regardless of definition) would likely not have adequate expertise for undertaking policy administration. However, some definitions of community include entities that could effectively and efficiently collect the revenue needed to pay for a community policy through special assessments, property taxes, and other means.
- **Confirming Compliance with Mandatory Purchase Requirements**
Currently, the mandatory purchase is only for the amount of a federally backed mortgage. A bundled community-based policy that provides a minimum required coverage would need to maintain some aspect of individual coverage and monitoring, which could be administratively burdensome. CBFI could cover all structures to a defined limit. Another alternative could be a CBFI that provides a set base coverage amount.
- **Pricing Expertise, Including Valuation of Mitigation Measures**
If private insurers were to underwrite risks associated with a CBFI policy, then they would want to price them according to actuarial principles. Private insurers thus would have to possess or acquire the information and expertise to price community-based policies to reflect the risk underwritten and the savings expected from mitigation measures. If the NFIP were to assume the risk of community-based policies, then presumably it would also assume the function of pricing these policies to account for the savings expected from mitigation measures. FEMA has expertise in setting premium costs based on flood risk and would have to work with communities to communicate individual property coverage costs bundled into a community policy.

References

Anderson, D. R. 1974. The National Flood Insurance Program: Problems and potential. Journal of Risk and Insurance 41(4):579-599.

Asche, E. 2014. Understanding the Value of Mitigation: Evidence from FEMA's Community Rating System Program and Elsewhere. Homeland Security, Studies and Analyses Institute. White Paper. Prepared for the Department of Homeland Security Science and Technology Directorate.

ASFPM (Association of State Floodplain Managers). 2013. Flood Mapping for the Nation: A Costs Analysis for the Nation's Flood Map Inventory. Association of State Floodplain Managers. Available at http://www.floods.org/ace-files/documentlibrary/2012_NFIP_Reform/Flood_Mapping_for_the_Nation_ASFPM_Report_3-1-2013.pdf [accessed May 19, 2015].

Berke, P., W. Lyles, and G. Smith. 2014. Impacts of federal and state hazard mitigation policies on local land use policy. Journal of Planning Education and Research 34(1):60-76.

Brody, S. D., S. Zahran, W. E. Highfield, S. P. Bernhardt, and A. Vedlitz. 2009. Policy learning for flood mitigation: A longitudinal assessment of the Community Rating System in Florida. Risk Analysis 29:912-929.

Brody, S. D., J. E. Kang, and S. Bernhardt. 2010. Identifying factors influencing flood mitigation at the local level in Texas and Florida: The role of organizational capacity. Natural Hazards 52:167-184.

Burby, R. 2006. Hurricane Katrina and the Paradoxes of Government Disaster Policy: Bringing about Wise Governmental Decisions for Hazardous Areas. Annals of the American Academy of Political and Social Science 604(March):171-191.

Coase, R. H. 1960. The problem of social cost. Journal of Law and Economics 3(1):1-44.

CRS (U.S. Congressional Research Service). 2012. The National Flood Insurance Program: Background, Challenges, and Financial Status. R40650. Available at http://www.washingtonpost.com/wp-srv/business/documents/health-science-NFIP-123110.pdf [accessed May 19, 2015].

CRS. 2013. The National Flood Insurance Program: Status and Remaining Issues for Congress. R42850. Available at https://fas.org/sgp/crs/misc/R42850.pdf [accessed May 19, 2015].

DePue, M., C. Carlton, D. Maurstad, and S. F. Wright. 2014. Community-Based Flood Insurance: Impacts on the Flood Hazard Management Cycle (PDF document). ASFPM 2014 Conference. Available at http://www.floods.org/Files/Conf2014_ppts/G3_DePue.pdf [accessed May 20, 2015].

Dixon, L., N. Clancy, S. A. Seabury, and A. Overton. 2006. The National Flood Insurance Program's Market Penetration Rate: Estimates and Policy Implications. Santa Monica, CA: RAND Corporation.

Dixon, L., N. Clancy, B. Bender, A. Kofner, D. Manheim, and L. Zakaras. 2013. Flood Insurance in New York City Following Hurricane Sandy. Santa Monica, CA: RAND Corporation.

FEMA (U.S. Federal Emergency Management Agency). 2002. National Flood Insurance Program: Program Description. Available at http://www.fema.gov/media-library-data/20130726-1447-20490-2156/nfipdescrip_1_.pdf [accessed May 20, 2015].

FEMA. 2005a. NFIP Floodplain Management Requirements. A Study Guide and Desk Reference for Local Officials. FEMA 480. Washington, DC: FEMA.

FEMA. 2005b. Joining the National Flood Insurance Program. FEMA 496. Washington, DC: FEMA.

FEMA. 2009. Severe Repetitive Loss Property Locations in FEMA Regions IV and VI. Available at http://www.fema.gov/media-library-data/20130726-1709-25045-4851/2_severerepetetiveloss.pdf [accessed May 29, 2015].

FEMA. 2011a. Testimony of Craig Fugate, Administrator of Federal Emergency Management Agency before the Senate Committee on Banking, Housing, and Urban Affairs, National Flood Insurance Program Reform, June 9, 2011, pp. 2.

FEMA. 2011b. NFIP Reform: Phase III Report. Washington, DC: FEMA. Available at http://www.fema.gov/pdf/business/nfip/nfip_reform_phase_III_report.pdf [accessed May 20, 2015].

FEMA. 2014a. Community Rating System Fact Sheet. Washington, DC: FEMA. Available at https://www.fema.gov/national-flood-insurance-program-community-rating-system [accessed February 2, 2015].

FEMA. 2014b. Cooperating Technical Partners Program. Available at https://www.fema.gov/cooperating-technical-partners-program [accessed on June 4, 2015].

FEMA. 2014c. FY 2014 Flood Mitigation Assistance Grant Program. Factsheet. Washington, DC: FEMA. Available at http://www.ak-prepared.com/grants/Documents/FY14%20FMA%20Fact% 20Sheet%20final%204-3-14_final.pdf [accessed April 29, 2015].

FEMA. 2015a. FloodSmart.gov: The Official Site of the NFIP. Available at https://www.floodsmart.gov/floodsmart/ [accessed February 3, 2015].

FEMA. 2015b. FEMA Flood Insurance Statistics. Available at https://www.fema.gov/total-coverage-calendar-year [accessed June 8, 2015].

FEMA. 2015c. Cooperating Technical Partners: Frequently Asked Questions. Available at https://www.fema.gov/cooperating-technical-partners-program/cooperating-technical-partners-frequently-asked-questions. [accessed June 16, 2015].

FEMA. 2015d. FY2015 Flood Mitigation Assistance. Fact Sheet. Available at http://www.fema.gov/media-library-data/1432847398289-878c470e718239eedcaadc8d52ea1823/FMAFactSheetFY2015.pdf [accessed on July 07, 2015].

FEMA. 2015e. Definitions. http://www.fema.gov/national-flood-insurance-program/definitions [accessed May 27, 2015].

Galloway, G. E. 1995. Learning from the Mississippi Flood of 1993: Impacts, Management Issues, and Areas for Research. U.S.-Italy Research Workshop on the Hydrometeorology, Impacts, and Management of Extreme Floods Perugia (Italy). Available at http://www.engr.colostate.edu/ ce/facultystaff/salas/us-italy/papers/12galloway.pdf [accessed May 27, 2015].

GAO (U.S. Government Accountability Office). 2010. FEMA Flood Maps: Some Standards and Processes in Place to Promote Map Accuracy and Outreach, But Opportunities Exist to Address Implementation Challenges. Washington, DC: GAO.

GAO. 2013. Flood Insurance: More Information Needed on Subsidized Properties. Report to Congressional Committees. Washington, DC: GAO.

GAO. 2014a. Overview of GAO's Past Work on the National Flood Insurance Program. Briefing to the Subcommittee on Housing and Insurance, Committee on Financial Services, House of Representatives. Washington, DC: GAO.

GAO. 2014b. Flood Insurance Foregone premiums. Washington, DC: GAO.

GAO. 2014c. Flood Insurance: Strategies for Increasing Private Sector Involvement. Washington, DC: GAO.

GAO. 2015. Flood Insurance: Status of FEMA's Implementation of the Biggert-Waters Act, as Amended. Washington, DC: GAO.

Harrington, S. E., and G. R. Niehaus. 2004. Risk Management and Insurance. Boston, MA: Irwin.

Hayes, T. L., and D. A. Neal. 2011. National Flood Insurance Program: Actuarial Rate Review: In Support of the Recommended Oct 1, 2011, Rate and Rule Changes. Washington, DC: FEMA.

Illinois Energy Management Agency. 2014. Disster Mitigation Act of 2000. Available at https://www.illinois.gov/iema/Mitigation/Pages/Planning.aspx [accessed May 25, 2015].

Kahneman, D. 2011. Thinking, Fast and Thinking Slow. New York: Farrar, Straus and Giroux.

Keybridge Research. 2011. Reforming the National Flood Insurance Program: A Qualitative Analysis of Policy Alternatives. Prepared for FEMA NFIP Reform Working Group.

Kousky, C., and L. Shabman. 2014. Pricing Flood Insurance. Washington, DC: Resource For the Future.

Kundzewicz, Z. W., S. Kanae, S. I. Seneviratne, J. Handmer, N. Nicholls, P. Peduzzi, R. Mechler, L. M. Bouwer, N. Arnell, K. Mach, R. Muir-Wood, G. R. Brakenbridge, W. Kron, G. Benito, Y. Honda, K. Takahashi, and B. Sherstyukov. 2014. Flood risk and climate change: Global and regional perspectives. Hydrological Sciences Journal 59(1):1-28.

Kunreuther, H., and E. Michel-Kerjan. 2013a. Implementing the National Flood Insurance Reform Act in a New Era of Catastrophes. Issue Brief. Penn Wharton Public Policy Initiative (1)9. Available at http://opim.wharton.upenn.edu/risk/library/2013WPPI_NFIP-brief.pdf [accessed April 22, 2015].

Kunreuther, H., and E. Michel-Kerjan. 2013b. A Proposal for Insuring Public Facilities and Infrastructure Against Disaster Losses. The Huffington Post Business. Available at http://opim.wharton.upenn.edu/risk/library/oped2013Oct29_HuffPost_InsuringPublicFacilities.pdf [accessed April 20, 2015].

Kunreuther, H., S. Gupta, V. Bosetti, R. Cooke, V. Dutt, M. Ha-Duong, H. Held, J. Llanes-Regueiro, A. Patt, E. Shittu, and E. Weber. 2014. Integrated Risk and Uncertainty Assessment of Climate Change Response Policies. In: Climate Change 2014: Mitigation of Climate Change. Contribution of Working Group III to the Fifth Assessment Report of the Intergovernmental Panel on Climate Change [Edenhofer, O., R. Pichs-Madruga, Y. Sokona, E. Farahani, S. Kadner, K. Seyboth, A. Adler, I. Baum, S. Brunner, P. Eickemeier, B. Kriemann, J. Savolainen, S. Schlömer, C. von Stechow, T. Zwickel and J.C. Minx (eds.)]. Cambridge University Press, Cambridge, United Kingdom and New York, NY, USA.

Langbein, W. B. 1953. Flood Insurance. Land Economics (29)4:232-330.

Lee, A. 2012. Sandy Catches Northeasterners without Flood Coverage. The Sun Herald. Available at http://www.weather.com/news/sandy-northeasterners-no-insurance-201211 [accessed April 22, 2015].

Lyles, W., P. Berke, and G. Smith. 2014. A comparison of local hazard mitigation plan quality in six states, USA. Landscape and Urban Planning 122:89-99.

Michel-Kerjan, E., S. Lemoyne de Forges, and H. Kunreuther. 2012. Policy tenure under the U.S. National Flood Insurance Program. Risk Analysis 32(4):644-658.

Milly, P. C. D., J. Betancourt, M. Falkenmark, R. M. Hirsch, Z. W. Kundzewicz, D. P. Lettenmaier, and R. J. Stouffer. 2008. Stationarity is dead: Whither water management? Science 319:513-514.

New Orleans Times Picayune. 2013. St. Charles town hall meeting on FEMA flood maps draws huge crowds eager to voice concerns. April 16.

NRC (National Research Council). 2013. Levees and the National Flood Insurance Program: Improving Policies and Practices. Washington, DC: National Academies Press.

NRC. 2015. Affordability of National Flood Insurance Program Premiums—Report 1. Washington, DC: National Academies Press.

Peterson, T. C., R. R. Heim Jr., R. Hirsch, D. P. Kaiser, H. Brooks, N. S. Diffenbaugh, R. M. Dole, J. P. Giovannettone, K. Guirguis, T. R. Karl, R. W. Katz, K. Kunkel, D. Lattenmaier, G. J. McCabe, C. J. Paciorek, K. R. Ryberg, S. Schubert, V. B. S. Silva, B. C. Stewart, A. V. Vecchia, G. Villarini, R. S. Vose, J. Walsh, M. Wehner, D. Wolock, K. Wolter, C. A. Woodhouse, and D. Wuebbles. 2013. Monitoring and Understanding Changes in Heat Waves, Cold Waves, Floods and Droughts in the United States: State of knowledge. Bulletin of the American Meteorological Society 94(June):821-834.

Pratt, J. W., and R. J. Zeckhauser. 1985. Principals and Agents: The Structure of Business. Boston, MA: Harvard Business School Press.

Rejda, G. E., and M. McNamara. 2014. Principles of Risk Management and Insurance. 12th Edition. Upper Saddle River, NJ: Prentice Hall.

Staff, Subcommittee on Economic Development, Public Buildings, and Emergency Management. 2015. Committee Hearing on "Rebuilding after the Storm: Lessening Impacts and Speeding Recovery." Committee on Transportation and Infrastructure, U.S. House of Representatives. Washington, DC.

Stocker, T. F., D. Qin, G. K. Plattner, M. Tignor, S. K. Allen, J. Boschung, A. Nauels, Y. Xia, V. Bex, and P. M. Medgley (eds.). 2013. Climate Change 2013: The Physical Science Basis. Working Group I Contribution to the Fifth Assessment Report of the Intergovernmental Panel on Climate Change-Abstract for decision makers: (Groupe d'expers intergovernmental sur l'evolution du climat/Intergovernment Panel on Climate Change-IPCC, C/O World Meteorological Organization, 7 bis Avenue de la Paix, CP 2300 CH-1211. Geneva 2 (Switzerland).

Task Force on Federal Flood Control Policy. 1966. A Unified Program for Managing Flood Losses. Washington, DC: U.S. Government Printing Office.

Tiebout, C. M. 1956. A pure theory of local expenditures. Journal of Political Economy 64(5):416-424.

Tversky, A., and D. Kahneman. 1974. judgment under uncertainty, heuristics and biases. Science 185(4157):1124-1131.

Van Zandt, S., W.G. Peacock, D. Henry, H. Grover, W. Highfield, and S. Brody. 2012. Mapping Social vulnerability to enhance housing and neighborhood resilience. Housing Policy Debate 22(1):29-55.

Zhao, W., H. Kunreuther, and J. Czajkowski. 2015. Affordability of the National Flood Insurance Program: A Case Study of Charleston County, South Carolina, Wharton Risk Management Center Working Paper #2015-03. Available at http://opim.wharton.upenn.edu/risk/library/WP201503_Affordability-of-Flood-Insurance-South-Carolina.pdf [accessed April 19, 2015].

Appendix A

Guest Speakers at Committee Meetings

Federal Agencies
Vincent Brown, FEMA, Arlington, VA
Bill Lesser, FEMA, Washington, DC
Andy Neal, FEMA, Arlington, VA
Brian Willsey, FEMA, Arlington, VA

State, City, and Regional Entities
Katherine Greig, Mayor's Office, New York City, NY
André McDonald, Fort Bend Flood Management Association, Fort Bend, TX
Bill Nechaman, New York State Department of Environmental Conservation, Albany, NY

Private Insurance Sector
John Hair, National Association of Mutual Insurance Companies, Washington, DC
Bob Sokolove, Bank of America, Charlotte, NC

Nonprofit Organizations
Len Shabman, Resources for the Future, Washington, DC
Carolyn Kousky, Resources for the Future, Washington, DC

Appendix B

Technical Discussion of the Responsibility for Insurance Is Irrelevant (RII) Proposition

For the sake of discussion and illustration, assume there are two actors, the community, c, and its residents, r. The analysis will be conducted using two residents, r_1 and r_2. The community takes actions, a, that diminish expected flood damages to the residents, d_1 and d_2, and to the community's infrastructure, d_c. Such actions could be building a dike or restoring a marsh. The community could also restrict residents from building or living in certain areas, or impose restrictions on residences, such as requiring that houses be built with pilings. For now, leave aside risk aversion for the individuals, as well as concerns about aggregate risk for the community and the insurer. Posit actuarially fair insurance, which enables one to expect flood damages as a measure of loss. The goal of these simplifications is to enable us to get our thinking straight in the simplest case.

The individuals take measures, m_1 and m_2, which reduce their personal expected flood losses but not any other party's losses. Posit that both the action of the community and the measures of the residents are calibrated in dollar terms. The objective is to have the combination of a and m_1 and m_2 that minimizes total flood costs (TFCs), as comprised by expenditures by both the residents and the community plus the damages to the residents and the community's infrastructure.

$$\text{Total Flood Costs} = a + m_1 + m_2 + d_1 + d_2 + d_c \quad (1)$$[1]

[1] If the community were to restrict locations or impose building requirements, then the tally of "damages" would also include the costs to the individuals in direct dollars or willingness-to-pay for meeting those impositions.

Posit, as is usually assumed and is highly likely to be true, that there are diminishing returns to a and to m_1 and m_2 in reducing damages over the relevant range of values.[2] Damages to a resident are limited by her measure plus the action of the community.[3] Damages to the community infrastructure are limited solely by the community's expenditure. Thus,

$$d_1 = f(m_1, a)\ ;\ d_2 = g(m_2, a);\ and\ d_c = h(a) \qquad (2)$$

The different damage functions for residents 1 and 2 arise because they reside in different places and have different structures.

The goal, to reiterate, is to find those values a^*, m_1^*, and m_2^* that minimize (1) given the production functions (2). Thus, substitute the values in (2) for d_1, d_2, and d_c and then substitute them into equation (1). Then take derivatives with respect to each of the three choice variables, a, m_1 and m_2, and set them equal to 0. Let the derivatives of the f, g, and h functions with respect to their arguments be denoted by subscripts of those arguments. The efficiency conditions are

For a

$$f_a + g_a + h_a + 1 = 0;\ for\ m_1\ f_{m1} + 1 = 0;\ and\ g_{m2} + 1 = 0 \qquad (3)$$

Subtracting 1 from each side of the three equations in (3) gives us more easily interpreted conditions:

For a

$$f_a + g_a + h_a = -1;\ for\ m_1\ f_{m1} = -1;\ and\ g_{m2} = -1 \qquad (4)$$

The interpretation of the efficiency conditions in (4) is straightforward. The community should continue to spend dollars until the sum of the reduction in expected damages to the two residents and itself is just \$1. The residents should only be concerned with themselves and should spend so \$1 of m reduces damages by \$1. As long as the community and the residents adjust to each other's actions, they need not coordinate in any way.

The outcome above will be achieved if the community takes its residents' damages fully into account, as it should, and if the residents optimize for themselves.

Now consider insurance provided by the federal government to the

---

[2] There may be increasing returns over some range. For example, building half a dike may not reduce risks by much. However, the relevant range is beyond such a point; it begins where returns start to diminish.

[3] Consider a one-period model, but implicitly are taking account of multiple periods. Thus, one damage-reducing measure might be locating in an area with lesser flood risk.

residents on an actuarially fair basis. The insurance cost, k, for expected damages, d, will just equal d. Thus, substituting k for d in the analysis above, everything goes through as before. (Moreover, risk aversion is ruled out as a concern, given insurance.) The requirement for efficiency is that the community now has to take the residents' insurance costs into account. The residents themselves have to raise their m until the sum of $m + k$, what is now their total costs, is minimized.

If the residents are fully insured, then they no longer suffer financially from damages. Thus, the efficiency condition for the community is that it take account of the residents' insurance costs, as it should. If insurance is only partial, then resident i will pick m_i to minimize $m_i + k_i + d_i$, the community will pick a to minimize the sum $d_c + [k_1 + d_1] + [m_2 + k_2 + d_2]$, and everything goes through as before.

Given the assumptions listed here in Appendix B, it matters not whether the community or the residents pay insurance premiums.

Appendix C

Committee Biographical Information

Henry J. Vaux, Jr., *Chair*, is professor emeritus of Resource Economics at the University of California in both Berkeley and Riverside. He is chair of the Rosenberg International Forum on Water Policy at University of California, Berkeley. Dr. Vaux is also the associate vice president emeritus of the University of California System, where he served as the chief operating officer for Agricultural and Natural Resource Programs and as director of California's Center for Water Resources. His principal research interests are the economics of water use, water quality, and water marketing. Prior to joining the University of California, he worked at the Office of Management and Budget and served on the staff of the National Water Commission. Dr. Vaux has served on the NRC committees on Assessment of Water Resources Research, Western Water Management, Ground Water Recharge, and Sustainable Underground Storage of Recoverable Water. He served chairman of the NRC Water Science and Technology Board from 1994-2001. He is a National Associate of the National Academies. Dr. Vaux holds an A.B. degree from University of California, Davis, in biological sciences and, an M.A. degree in natural resource administration. He holds an M.S. degree in economics and Ph.D. degree in natural resource economics from the University of Michigan.

Patricia Born is the Midyette Eminent Scholar of Insurance in the Department of Risk Management/Insurance, Real Estate and Legal Studies at Florida State University (FSU). She is director of the FSU Center for Insurance Research and is a research associate in the Florida Catastrophic Storm Risk Management Center, the Center for Innovative Collaboration in Medicine and Law, and the FSU Institute for Successful Longevity. She is also the director of the Risk Management/Insurance Ph.D. program in the College of Business and holds a courtesy appointment at the FSU College of Law. Her research interests include the regulation of insurance, medical malpractice, tort reform, health insurance, and the modeling and management of catastrophic risks. She has published in leading insurance academic journals including *Journal of Risk and Uncertainty*, *Journal of Risk and Insurance*, *Journal of Regulatory Economics*, *Columbia Business Law Review*, *Insurance: Mathematics and Economics*, and the *Journal of Business and Economic Statistics*. She is currently president of the American Risk and Insurance Association and chair of the Florida Panhandle District Long Term Care Ombudsman Council. Recent consulting clients include Florida Department of Transportation, TIAA-CREF, the Insurance Bureau of Canada, the National Association of Mutual Insurance Companies, and the Able Trust. Dr. Born holds her Ph.D. degree in economics from Duke University.

Jeffrey Czajkowski is a senior research fellow at the Wharton Risk Management and Decision Processes Center, serving as the Willis Research Network Fellow. His primary research fields are the economics of natural hazards and environmental economics where his research has focused on integrated modeling of direct economic losses associated with natural hazard event occurrences such as hurricanes, flood, and hail; quantifying risk-based insurance premiums related to catastrophic events; estimating the benefits associated with short- and long-term natural hazard preparation and mitigation activities; modeling and understanding economic decision making in the presence of a natural disaster; and economic valuation of environmental goods via revealed and stated preference techniques. Prior to his position at the Risk Center, Dr. Czajkowski was an assistant professor of economics at Austin College, Texas. Through September 2009, he was also an adjunct assistant research professor at the International Hurricane Research Center (IHRC) at Florida International University (FIU). During his tenure at FIU, he was an E.P.A fellow, as well as an FIU dissertation fellow. Preceding graduate school he worked as a research associate for Coopers & Lybrand Consulting and a vice-president for JP Morgan in New York City. Dr. Czajkowski holds a B.S. degree from Carnegie Mellon University, an M.S. degree in environmental and urban systems from FIU, and a Ph.D. degree in economics from FIU.

Lloyd Dixon is a senior economist and director of the Center for Catastrophic Risk Management and Compensation at the RAND Corporation (Santa Monica, California) where he specializes in insurance, compensation, and liability issues. Dr. Dixon has conducted a number of studies on flood and wind insurance markets. His studies on the National Flood Insurance Program have provided more accurate estimates of the take-up rate for flood insurance and the compliance rate with the program's mandatory purchase requirement and examined the role of the private market in underwriting residential flood insurance. Dr. Dixon served on the New York State 2100 Commission set up by Governor Cuomo in the wake of Hurricane Sandy. Dr. Dixon has also authored several studies and has spoken widely on government intervention in the market for terrorism insurance. Work is currently under way on the effects of TRIA's potential expiration on workers' compensation insurance markets. He has also assessed the roles played by insurance, charity, tort, and government programs in providing assistance and compensation to individuals and businesses affected by the September 11 attacks. Dr. Dixon holds a B.S. degree in engineering and a B.A. degree in political science from Stanford University and a Ph.D. degree in economics from University of California, Berkeley.

Robert M. Hirsch is a research hydrologist at the U.S. Geological Survey (USGS) in Reston, VA. The focus of his research is the description and understanding of long-term variability and change in surface-water quality and streamflow. From 1994 through May 2008, he served as the chief hydrologist of the USGS. In this capacity, Dr. Hirsch was responsible for all USGS water science programs. From 2003 to 2010 he served as the cochair of the Subcommittee on Water Availability and Quality of the Committee on Environment and Natural Resources of the National Science and Technology Council. Dr. Hirsch has received numerous honors from the federal government and from nongovernmental organizations, including the 2006 American Water Resources Association's William C. Ackermann Medal for Excellence in Water Management, and has twice been conferred the rank of Meritorious Senior Executive by the President of the United States. He is a recipient of the USGS "Eugene M. Shoemaker Award for Lifetime Achievement in Communications." He is coauthor of the textbook *Statistical Methods in Water Resources*. Dr. Hirsch is a fellow of the American Association for the Advancement of Science and an active member of the American Geophysical Union and the American Water Resources Association. He has testified before congressional committees on many occasions and presented keynote addresses at many water-related meetings in the United States and other countries. Dr. Hirsch holds a B.A. degree in geology from Earlham College, an M.S. degree in geology from the University of Washington, and a Ph.D. degree from The Johns Hopkins University.

Roger Kasperson is a research professor and distinguished scientist at the Graduate School of Geography, Clark University. Prior to Clark University, Dr. Kasperson taught at the University of Connecticut and Michigan State

University. He has written widely on issues of risk analysis, risk communication, global environmental change, risk and ethics, and environmental policy. Dr. Kasperson is a member of the National Academy of Sciences and the American Academy of Arts and Sciences. He is a recipient of the 2006 Distinguished Achievement Award of the Society for Risk Analysis. He has been a consultant or advisor to numerous public and private agencies on energy and environmental issues and served on various committees of the National Research Council and the Council of the Society for Risk Analysis. From 1992 to 1996 he chaired the International Geographical Union Commission on Critical Situations/Regions in Environmental Change. He was vice president for academic affairs at Clark University from 1993 to 1996, and in 1999 he was elected director of the Stockholm Environment Institute, a post he held through 2004. His current research and teaching include vulnerability and resilience to global climate change, initiatives to integrate sustainability to environmental policies, and the management of high uncertainty risks. Dr. Kasperson holds a B.A. degree from Clark University. Both his M.A. and Ph.D. degrees were from the University of Chicago.

Robert Klein is an associate professor of risk management and insurance and director for the Center of Risk Management and Insurance Research at Georgia State University in Atlanta. Dr. Klein is a leading expert on insurance regulation and markets with 30 years of experience as a regulator and an academic researcher. He has published extensively on various topics in insurance and its regulation, including the structure and performance of insurance markets, solvency regulation, monitoring competition, price regulation, catastrophe risk, homeowners insurance, urban insurance issues, workers compensation, life insurance, and international insurance regulation. He has also testified frequently at legislative and regulatory hearings on significant issues affecting insurance consumers and the industry. Prior to joining Georgia State University in September 1996, Dr. Klein was the director of research and chief economist for the National Association of Insurance Commissioners. He has also served as staff economist for the insurance department and state legislature in Michigan. Dr. Klein is a Sloan Fellow at the Financial Institutions Center at the Wharton School of Business. He has served on the Board of Directors for the American Risk and Insurance Association and currently serves on the editorial boards for the *Journal of Insurance Regulation and Risk Management and Insurance Review*. He is a member of the American Economic Association, the American Risk and Insurance Association, and the Southern Risk and Insurance Association. Dr. Klein holds his B.A., M.A., and Ph.D. degrees in economics from Michigan State University.

Sandra Knight is a senior research engineer in the Department of Civil and Environmental Engineering at the University of Maryland where she works in the development of water policy, disaster resilience, and flood risk management initiatives to support the Center for Disaster Resilience. Additionally, she is founder and president of WaterWonks LLC in Washington, DC. Dr. Knight finished her federal career as the deputy associate administrator for mitigation at the Federal Emergency Management Agency, responsible for the nation's floodplain mapping, management and mitigation grants supporting the National Flood Insurance Program, environmental compliance for the agency, and oversight of the National Dam Safety Program. At National Oceanic and Atmospheric Administration (NOAA), she was responsible for the development of policies and strategies to ensure scientific excellence and improved performance of NOAA's research portfolio. Prior to that, she spent 26 years with the U.S. Army Corps of Engineers. Her last position with the U.S. Army Corps of Engineers was as technical director for navigation research. She is a registered professional engineer in Tennessee, a Diplomate Water Resource Engineer, and a Diplomate Navigation Engineer. She is a member of the American Society of Civil Engineers, the American Meteorological Society, the Society of Women Engineers, Sigma Xi, and a fellow for PIANC. Dr. Knight holds a B.S. degree from Memphis State University, an M.S. degree from Mississippi State University, and Ph.D. degree from University of Memphis—all in civil engineering.

David I. Maurstad is a director and senior vice-president with Optimal Solutions and Technologies, Inc., a provider of management consulting, integrated information technology, engineering services, and business process outsourcing in Washington, DC. Mr. Maurstad previously served as director of water policy and planning for a nationally recognized engineering firm specializing in flood mapping and floodplain management. He has more than 30 years of leadership experience with both the private insurance industry and federal, state, and local government. In June 2004, he was appointed by President George W. Bush to provide leadership for some of the nation's leading multi-hazard risk reduction programs. In this role he was the federal insurance administrator charged with the overall management of FEMA's National Flood Insurance Program. He previously served as the director of FEMA Region VIII (2001-2004) coordinating federal, state, tribal, and local management of emergencies through planning, preparedness, mitigation, response, and recovery. Mr. Maurstad is a former lieutenant governor and state senator of Nebraska and served as mayor of Beatrice, Nebraska. Mr. Maurstad holds a B.S. degree in business administration and M.B.A. degree from the University of Nebraska.

Sally McConkey is the head of the Coordinated Hazard Assessment and

Mapping Program (CHAMP) at the Illinois State Water Survey (ISWS). The 25 member CHAMP staff is composed of engineers, geographic information (GIS) specialists, and outreach staff. Ms. McConkey has been with ISWS since 1984. Ms. McConkey has also served on the Illinois Association for Floodplain and Stormwater Management (IAFSM) board in a variety of positions including vice chair and chair. She served as cochair of the Mapping and Engineering Standards Policy Committee of the Association of State Floodplain Managers (ASFPM). Ms. McConkey also served as vice-chair and chair of ASFPM. Ms. McConkey is currently serving on the Technical Mapping Advisory Council to FEMA. She is a registered professional engineer in the State of Illinois, a certified floodplain manager, and a Diplomate, Water Resources Engineer. Ms. McConkey holds a B.A. degree in mathematics, a B.S. degree in theoretical and applied mechanics, and an M.S. degree in civil engineering from the University of Illinois.

Tommy Wright has been chief of the Center for Statistical Research and Methodology (formerly Statistical Research Division), U.S. Census Bureau, since January 1996 and an adjunct member of the faculty at Georgetown University since 2009. From 1979 to 1996, he was a research staff member of the Mathematical Sciences Section at Oak Ridge National Laboratory. Author of one book, editor of another, and author of more than 40 papers in statistics and mathematics journals, his research interests have focused on probability sampling and estimation and elementary applied probability and combinatorics. He has more than 30 years of undergraduate/graduate teaching experience in statistics and mathematics. His contributions and service have led to professional recognition: (1) elected member, International Statistics Institute (1989) and (2) fellow, American Statistical Association (1995). Dr. Wright holds a B.S. degree in mathematics from Knoxville College and an M.S. degree in mathematics from the University of Tennessee. He holds an M.S. and Ph.D. degrees in statistics from The Ohio State University.

Richard Zeckhauser is the Frank P. Ramsey Professor of Political Economy at the Kennedy School, Harvard University. He is an elected fellow of the Econometric Society, the Institute of Medicine, and the American Academy of Arts and Sciences. In 2014, he was named a Distinguished Fellow of the American Economic Association. His contributions to decision theory and behavioral economics include the concepts of quality-adjusted life years (QALYs), status quo bias, betrayal aversion, and ignorance (states of the world unknown) as a complement to the categories of risk and uncertainty. Many of his policy investigations explore ways to promote the health of human beings, to help markets work more effectively, and to foster informed and appropriate choices by individuals and government agencies.

He has published more than 280 articles. His recent coauthored books are *The Patron's Payoff: Conspicuous Commissions in Italian Renaissance Art* (2008) and *Collaborative Governance: Private Roles for Public Goals* (2011). Dr. Zeckhauser holds a Ph.D degree from Harvard College (summa cum laude).